Green Land, Brown Land, Black Land

Green Land, Brown Land, Black Land

An Environmental History of Africa,
1800–1990

James C. McCann

HEINEMANN
Portsmouth, NH

JAMES CURREY
Oxford

Heinemann
A division of Reed Elsevier Inc.
361 Hanover Street
Portsmouth, NH 03801–3912

James Currey Ltd.
73 Botley Road
Oxford OX2 0BS
United Kingdom

ISBN 0–325–00096–4 (Heinemann)
ISBN 0–85255–774–4 (James Currey)

Library of Congress Cataloging-in-Publication Data

McCann, James, 1950–
    Green land, brown land, black land : an environmental history of
  Africa, 1800–1990 / James C. McCann.
      p. cm.
    Includes bibliographical references and index.
    ISBN 0–325–00096–4 (alk. paper)
    1. Africa—Environmental conditions—History.   I. Title.
  GE160.A35M36   1999
  363.7'0096—dc21   98–35565

British Library Cataloguing in Publication Data Available

ISBN 0–85255–774–4 (James Currey)

Cover illustration: Lesotho landscape by Leetsang Ncheke (March 1976). Courtesy of David Ambrose.

Printed in the United States of America on acid-free paper.

Docutech T & C 2007

To My Colleagues and Friends in the
African Studies Center
at Boston University, 1984–1998.

Contents

viii

Contents

List of Figures

Preface and Acknowledgments

This book had its genesis in spring 1995 in a hallway of the African Studies Center at Boston University when my colleague Jean Hay approached me with an idea she had for a new Heinemann title. She wanted me to take a new course I was developing on African environmental history and work it into a published form, a new sort of monograph that blended my research interests with a wider thematic scope. It seemed, she said, a logical extension of my teaching, my research interests in agriculture and ecology, and Heinemann's interest in a book that addressed environment as a coherent thread of African history.

It took me a couple of days to realize what a good idea this was and to warm to the potential for publishing such a title. It took me a couple more months to realize how much work this project would demand and what its risks would be. A great deal of interesting work was just then coming out from a new generation of scholars of the environment in African history, geography, and anthropology that I wanted to draw together in some comprehensive way. I would need to make serious choices about subthemes and geography, and find a way to get myself onto some key historical landscapes. The book could use some of my own research but would have to rely most heavily on my reading of others' work. This book is a result of that process of meeting people, exchanging ideas, and making choices about to deal with Africa's geographic and historical diversity. The danger, of course, is that works of synthesis invariably invite a critical eye from specialists whose depth of knowledge and experience in a particular place necessarily exceeds the author's. As author I stand on the shoulders of many others but ultimately I must take responsibility for the choices I have made here.

Nevertheless, I have tried as much as possible to expand on my own field experience, to visit and interpret firsthand the landscapes I write about, and to consult people with long experience. Research and travel for this book was made possible by a Fulbright-Hays Faculty Research Abroad grant for the summers of 1996 and 1997. Beginning with my first dissertation grant in 1980, I have been thrice blessed with Fulbright-Hays grants from the U.S. Department of Education's Office of International Education. John Paul and the staff there have offered extraordinary assistance and understanding. The Fulbright-Hays grant allowed me to travel to Ethiopia, Lesotho, South Africa, Ghana, and Britain. In each of those places I found people willing to share their research and experiences with me and to help me "read" historical landscapes in insightful ways.

In Ethiopia, I had invaluable help from the staff at the International Livestock Centre for Africa (now the International Livestock Resource Institute), especially Degefe Birru, Abate Tedla, and Mohamed Saleem. Helge Espe and Belachew Sobuku of Norwegian Redd Barna helped get me to sites otherwise unreachable because of local politics or swollen rivers. Dr. Mesresha Fetene of the Biology Department of Addis Ababa University and Tewolde Berhan Gebre Egziabher of the Ethiopian Environmental Protection Agency offered helpful introductions and insights. The staff of the National Herbarium provided timely assistance.

Lesotho was a new landscape for me, and I needed the local knowledge of Kate Showers, John and Judy Gay, Chaba Mokoko, David Ambrose Tumelo Tsikoane, David Hall, and David May to navigate my way through gray literature and mountain trails. Special thanks go to Stephan Gill of the Morija Museum and Archives, and to Albert Brutsch, who led me into the archives' valuable primary sources and photographic materials. There is no substitute for the excitement and insights of reading primary accounts on the very landscape being described.

In Bloemfontein (South Africa) I had special help and guidance from Leo Barnard and Geoff de Villiers of the University of the Free State. Chester Willnot, a fourth-generation farmer took time to describe the history of pasture and dryland farming in the Eastern Free

State. Marianne Botes of the National Museum and Steven Hundt of the Oliewenhuis Art Museum provided advice and access to valuable photographs and museum collections.

Ghana has, in my experience, the world's most hospitable national culture. For confirming that opinion, I owe special thanks to Dr. Kofi Baku, Fosuaba Mensah Banehene, Issac Owusu, Chuck Hutchison, Afriyie S. Twumasi, and Ben Dzah for leading me through libraries, markets, biology labs, forest canopy walks, and crop research stations.

During my stay in Britain I enjoyed the hospitality and insights of a superb group of scholars and staff at the Institute of Development Studies at the University of Sussex. Special thanks are due to James Fairhead, Melissa Leach, Ian Scoones, Grace Carswell, and Annette Sinclair. In London, David Anderson and Richard Rathbone have always offered key introductions and collegial support.

Claire Ivison and Catherine Lawrence of the SOAS Department of Geography drafted the book's maps with patience and care. I am also grateful to Hal Rothman and Richard Mingus of *Environmental History* for permission to use portions of Chapter 5 that had previously appeared in that journal. Valuable comments on that chapter also came from James Scott and the 1996 Fellows at the Program in Agrarian Studies at Yale University, Alan Taylor, and the members of the Center for Agricultural History at the University of California at Davis. Other colleagues who have offered comments and encouragement in this project include Jim Giblin, Nancy Jacobs, Greg Maddox, Tom Spear, and Shiferaw Bekele.

Back home in Boston there are special people who have long offered intellectual and emotional sustenance. At the African Studies Center that group has included, of course, Jean Hay, Joanne Hart, and Maddelena Goodwin. During the gestation of this book a set of graduate students have provided inspiration and ideas: Erik Gilbert, Heather Hoag, Kirk Hoppe, Tom Johnson, Yusefu Lawi, Sarah Phillips, and Tumelo Tsikoane. Dennis Berkey, Dean and now Provost at Boston University, has been a constant source of support for scholarship on Africa.

Again, as before, members of my family have been constant companions whether I am at home or away. Sandi, Libby, and Martha are my sources of laughter and inspiration. They have also been stalwart travel partners from Cape Point to Tsitsikama and beyond.

Chapter 1

Introduction

Africa's environmental history is written on its landscapes. While the continent's geomorphology—mountains, river valleys, coastlines—changes at a pace imperceptible to human generations, the shades and textures of its soils, forests, vegetation, and human settlement reflect its history in a way more profound and ubiquitous than politics, economics, or even colonial rule. Africa's modern landscapes vary widely: from the asphalt and corrugated iron to the glass and steel of sprawling new cities to rural fields covered increasingly with New World crops (maize, cassava, cocoa); from semiarid savanna woodland to rain forests; from coastal mangrove swamps to highland euphorbia plateaux. Moreover, Africa's patchwork of landscapes continues to change, perhaps at a rate unprecedented in its history, as exotic species and new ideas arrive and join a changing mix of plant, animal, and viral inhabitants.

If Africa's physical landscapes have been a canvas, then nature's palette of colors and textures has been the action of climate, the life cycles of vegetation, and the action of water. These landscapes change not only in a linear fashion over time, but also within the year in the form of seasons. Wet summers and dry winters transform savanna and open woodland from golden to green and river beds from dry or sluggish rivulets to grand waterways or fast-moving torrents.

Like water, fire was a seasonal force, sometimes human induced, sometimes the result of the serendipity of a lightning strike. Fire was thus both a tool and a natural force that transformed landscapes for a few months or over an environmental epoch. In the short term a grass fire removed dried vegetation and allowed new shoots to emerge at first rains. Fire and the ashes of wood and grass changed soil Ph and released phosphorous. Over the long term fire encouraged the spread of fire-resistant tree species and confined other trees to protected areas.

The seasonal metamorphosis of vegetation also draws to it Africa's astonishing menagerie of fauna, lives that depend directly or indirectly on the food value, soil effects, and disease ecology within the vegetative cover. This fauna includes large mammals, both domestic and wild, who feed directly on African grasses, as well as carnivores—feline, canine, and avian—who follow them as their food sources. Finally, another set of less visible predators inhabit Africa's landscapes. Disease organisms—protozoans, viruses, nematodes—bring epidemic and endemic disease to humans and animals. Insects—locusts, tsetse fly, black fly, ticks, and lice—move along with the natural movement of the seasons, vegetation, and temperature, pushing and directing life-forms in subtle ways.

Above all of these factors of environmental change shaping African landscapes have been the labor, tools, and ideas of Africa's human inhabitants. A fundamental leitmotif in this book is the premise that Africa's landscapes are anthropogenic, that is the product of human action. Thus, Africa's landscapes show the cumulative effects of specific human tools: hand hoes, oxplows, axes, machets, and human agents such as domestic livestock, fire, crops. The impact of these factors through most of the period covered in this book, therefore, also depends directly on demography, that is the effects of varying concentrations of human settlement on particular parts of African landscapes.

Snapshot views or images of long-term changes in Africa's environment depend on two scales: time and space. Africa's forests, soils, and animal populations have changed over time, but their collective effects on landscapes also depend on geographic scale. This book incorporates a number of possible perspectives on environmental change, ranging from a Landsat satellite's space-based camera to the view from an airliner breaking through cloud cover at 3000 meters; the field of vision from the edge of a farmer's field or a shepherd's perch. Each of these points of observation would capture a different scale but also a different point of view: the shepherd's concern for livestock's safety and food supply, the farmer's preference for an open, cropped field; and an international space agencies' overview of land use measured in 100 kilometer square blocks.

There is also a grander scale of economy that has increasingly affected Africa's physical environment over the course of the 1800–1990 period. While the forces that directly shape Africa's landscapes are local, they are increasingly conditioned indirectly by events and choices at a more global scale. That hierarchy of scale extends from the farmer's field to the regional marketplace to the Ministries of Mines, Agriculture, or Finance and to the offices of multilateral agencies in Washington, Rome, or Nairobi. Over the two centuries covered in this book the scale of human capacity to change Africa's environment has changed dramatically as technology and the growth of international commodity markets for coffee, cocoa, wood, and minerals have dominated local economies. In the late twentieth century cocoa merchants in Amsterdam arguably have more control than do local farmers over whether Ghana's region is covered with cocoa forests or fields of maize. The use of concrete for an apartment complex's roof in Kuwait may save a mangrove swamp in Tanzania's Rufigi delta.

This book seeks to explore the process of interaction between the physical world of plants, soils, climate, and animals with human action and response over the period 1800 to 1990. Environmental history rarely falls neatly into specific dates, nor can a history that encompassed all of Africa rest on a fixed bookend date. The benchmark of 1800 is thus an attempt to incorporate a precolonial past along with the full range of Africa's engagement with the industrial world economy, colonialism, and global economic change in the late twentieth century.

Environmental and landscape history is also, to a large degree, the history of ideas, perceptions, and prescriptions about what historical African cultures and colonial governments felt about how land should look. Their actions on the land reflected deeply rooted aesthetic traditions about natural and inhabited space and the social organization of technology and labor power to transform it.

In the late twentieth century Africa's environment has changed from what Africans and outside observers saw around them in 1800. There are fairly widespread beliefs that degradation rather than merely change has been a dominant theme: the alleged processes include deforestation, erosion, loss of soil fertility, increasing drought, and the loss of biodiversity. Media imagery and accounts of declining natural resources have dominated public perceptions of Africa. How accurate are these assertions of environmental decline? This book will test those arguments and illustrate the processes that shaped Africa's environmental history in the last two centuries of the millennium.

The environmental history of Africa is beginning only recently to take a coherent shape with the publication of a new generation of empirical, field-based studies that challenge firmly held assumptions about past patterns of land use by documenting past conditions of African natural resources and human response to environmental change. These studies have dispelled the myth of an African Eden lost and established a more workable hypothesis that Africa's historical landscapes are both historically and currently anthropogenic. At the same time, these studies have begun to reexamine the relevance of powerful and resilient theoretical paradigms—most notably those of Thomas Malthus and Ester Boserup—that have driven assumptions about the dynamics of population and natural resource use in past time.[1]

Parallel to the new research on Africa's historical landscapes has been a broader understanding of the role of narratives, stories about the environmental past that underlie public policy, shape the thinking of popular culture, and spur the development of scholarly debates on Africa. Allan Hoben from the perspective of development anthropology has described these ideas about African natural resource use as:

> historically grounded, culturally constructed paradigms that at once describe a problem and prescribe its solution. Many of the environmental narratives about Africa are rooted in a narrative that tells us how things were in an earlier time when people lived in harmony with nature, how human agency has altered that harmony, and of the calamities that will plague people and nature if dramatic action is not taken soon.[2]

Historian William Cronon points out that environmental history is particularly susceptible to powerful interpretive narratives because of the need to ascribe agency and order to human actions on the physical world. Scientific descriptions of physical processes can be obtuse, but narratives with human actors staged on the physical landscape provide a story line that can emphasize either folly or the triumph of human spirit, depending on the narrator's perspective on the nature of the past and the prescribed outcome. Cronon observes, for example, that the same physical evidence on the history of the American dust bowl can just as easily illustrate a story of human triumph over a hostile natural world as it can one of the perfidy of those who mismanaged their God-given resources.[3] For Africa as a whole and for the individual cases described in this book, these degradation narratives tend to be poorly informed about the African environmental past and to use ill-founded constructions of the past to foresee Malthusian futures.

Africa's physical size, the scale of its human mosaic, and its biological diversity defy both generalization and full coverage. Moreover, my own field experience is rooted in East Africa with only brief forays south and west. I have chosen, therefore, to illustrate rather than chronicle Africa's environmental history; to explain widespread patterns through examples; and to draw those cases from Africa south of the Sahara. This approach necessarily excludes North Africa, though it does not deny Africa's deep historical ties to the Mediterranean world. Part 1 of this book sets an historical context for Africa's environment in the 1800–1990 period by describing the ways in which Africa's precolonial past placed humans in nature. Chapter 2, "Africa's Physical World," outlines the physical building blocks of Africa's environment, including climate, soils, flora, fauna, and principles of demography. Chapter 3, "Environment and History in Africa," surveys the ecological context of historical change in several of Africa's most significant ancient civilizations. The goal here is not to offer environmental determinism as a simple guide to African history but to suggest environmental context as an important element of historical events and processes.

Part 2, "Africa's Historical Landscapes," examines the historical evidence for arguments (sometimes called "declensionist") that highlight the degradation of Africa's natural resources within the modern period. Chapter 4, "Desert Lands, Human Hands," explores the historical evidence for desertification and the role of human action in climate change drawn from West Africa and a new generation of research on the interaction between forest and savanna. Chapter 5, "A Tale of Two Forests," examines the historical evidence for allegations of decline in Ethiopia's forest cover over the past 150 years. Chapter 6, "Food in the Forest: Biodiversity, Food Systems, and Human Settlement in Ghana's Upper Guinea Forest," looks at the implications for biodiversity in the historical setting of Ghana's Upper Guinea forest. Chapter 7, "Soil Matters: Erosion and Empire in Greater Lesotho, 1830–1990," presents evidence for the history of land use, colonialism, and erosion in Lesotho and adjacent areas of the Free State in southern Africa. "Epilogue: Africa's Environnmental Future as Past" assesses the trends affecting Africa's natural resources and human settlement.

NOTES

1. For long-term studies of natural resource management in Africa, see Melissa Leach and Robin Mearns, *The Lie of the Land: Challenging Received Wisdom in African Environmental Change and Policy* (Oxford and Portsmouth, N.H., 1996); Kate B. Showers, "Soil Erosion in the Kingdom of Lesotho: Origins and Colonial Response, 1830s–1950s," *Journal of Southern African Studies* 15, no. 2 (1989): 263–86; Mary Tiffen, Michael Mortimer, and Francis Gichuki, *More People, Less Erosion: Environmental Recovery in Kenya* (Chichester, 1994).

2. Allan Hoben, "Paradigms and Politics: The Cultural Construction of Environmental Policy in Ethiopia." African Studies Center Working Papers No. 193, African Studies Center, Boston University, 1995; see also Allan Hoben, "Paradigms and Politics: The Cultural Construction of Environmental Policy in Ethiopia," *World Development* 23, no. 6 (1995): 1007–21.

3. William Cronon, "A Place for Stories: Nature, History, and Narrative," *The Journal of American History* 78 (March 1992): 1372–76.

PART I

Patterns of History

Chapter 2

Africa's Physical World

In his recent general history of Africa the distinguished Cambridge University historian John Iliffe described Africa's people as "the frontiersmen of mankind . . . who have colonized an especially hostile region of the world on behalf of the entire human race. That has been their chief contribution to history."[1] Iliffe goes on to identify the key themes of Africa's history as the expansion of population, the achievement of human coexistence with nature, and the building of enduring societies. For Iliffe, however, the overall theme that binds Africa's long, diverse history is humanity's struggle against a hostile natural world—more hostile he argues than what faced humankind in Eurasia or the Americas.

Choosing a theme of the human conquests of Africa's harsh physical endowments contrasts sharply with older images in literature and African nationalist politics that assert Africa's inherent wealth and tropical exuberance, that is, the richness of its vegetation and its wealth in minerals, both mythical (King Solomon's Mines, Sofala's El Dorado) and actual (the gold of the Witswatersrand, Sierra Leone's diamonds, and Zaire's copper). Indeed the exploitation of Africa's natural endowments by outsiders has been a key theme of both the continent's history and the way historians have written about it. Thus, whether a writer's leitmotif has been wealth or poverty, the environment has inevitably been an explicit or subtle subtheme that underlies Africa's historical development from the origins of humankind to modern history.

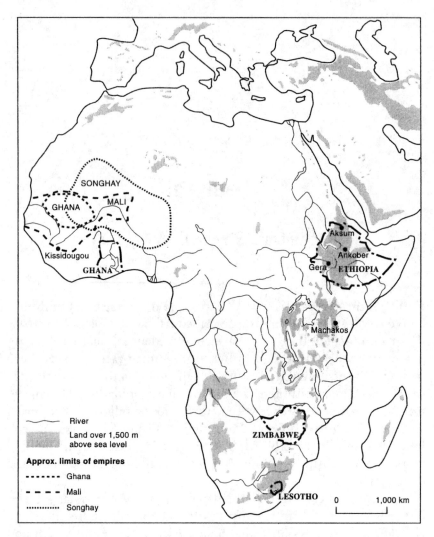

Figure 2.1   Map of Africa with Case Study Sites.

The aim of this chapter and the next is to examine a number of prominent themes of African history—state building, trade, population, and climate change—in the context of environmental history. The question of whether Africa's endowments of nature represent wealth or poverty is not the point, although John Iliffe's challenge to reconstruct the interactions of nature and humankind is a primary point of departure. The central question is: How does a knowledge and sensitivity to environmental issues provide a greater understanding of the defining movements and events of African history? This chapter will explore the role of nature and environment in a number of classic subjects of African history in the premodern period.

Africa has never been isolated intellectually, economically or ecologically. African history was and has been distinct and individual, but porous; the continent's interactions with wider worlds well before the modern period and colonial rule of the late nineteenth and early twentieth century were inextricably part of its historical dynamism. There have been a number of specific zones of interaction mediated by environmental factors that permitted flows of ideas, goods, and peoples into and out of Africa. These included the Mediterranean basin, the Red Sea/Nile Valley, the Indian Ocean, and, later, the Atlantic economy that bound Africa, Europe, and the New World to a complex but unequal exchange. Many or most of these avenues of exchange and communication were made possible by environmental conditions: seasonal winds (the Indian Ocean trade), oases in desert tracks (the trans-Saharan caravan trade), compatible climate patterns (the Mediterranean basin), or the exchange of new crop germplasm and disease organisms (the Atlantic system). With the arrival of the new "technology" of the camel in North Africa after the third century A.D., even the Sahara desert was an avenue of exchange, less a barrier than a permeable membrane that filtered contact and exchange between African and Mediterranean worlds.

OLD SOILS, HIGH LANDS

Africa is the oldest continent and the largest fragment of what was the land mass of Gondwanaland. Africa's soils and geomorphology show both its age and also more recent episodes of volcanic upheavals, depositing of sediments, and the action of water and wind across the land's surface. Altitudes range from 120 meters below sea level in the Danakil depression near the Red Sea to 5895 meters above sea level at the peak of Mt. Kilamanjaro. In northeast Africa geological domes formed from Ethiopia to Tanzania split open 750,000 years ago to create the Great Rift Valley; the Great Lakes formed when the Eastern Rift Dome lifted the earth's crust to form a series of long, deep lake basins stretching from south of the equator to include the Red Sea itself. The highlands in northeast Africa form what biologist Jonathan Kingdon calls a "fractured dome."[2] While only 4 percent of Africa's land mass is above 1500 meters, 50 percent of those highlands are in the Ethiopia/Eritrea region, giving the continent an overall tilt from northeast to southwest. Forty percent of Africa's land has a slope of more than 8 degrees, resulting in the movement and redeposition of soils by rain and river systems—erosion—not so much a recent crisis as a consistent and inexorable historical process.

From a satellite's view, however, the continent's major overall features are vast alluvial plains broken by the branching of old stream beds and living rivers. At ground level the textures and hues of soils are more visible. Africa's soils vary dramatically in color, chemistry, and structure, ranging from light, sandy arenosols to heavy, poorly drained black "cotton" soils (vertisols). Acid soils cover 18 percent of the continent while the red porous laterite soils that easily lose nutrients cover much more. Above all, however, soil conditions, erosion, and soil formation are all local phenomena with few generalizations possible. Erosion rates have been grossly exaggerated for places like the Ethiopian highlands where many observers have underestimated the historical effects of soil movement and new soil formation.[3] Africa's soils are not so much lost as they move and redeposit themselves.

For this study two issues are relevant: geomorphology and soil behavior. Geomorphology is, essentially, the shape of the land, since in Africa elevation and contours mediate the effects of climate, temperature, soil depth, and human settlement. Soils fit only loosely into chemical typologies since each location has a slightly different geological history, structure of organic admixture, and history of deposition. African soils have distinctive, localized personalities and continue to be actors in environmental history in their own right. Observing and describing soil behavior is a far more useful exercise than attempting to place it in a formal typology.

BIODIVERSITY: OLD AND NEW

Biodiversity may be defined simply as the number and variety of species of life and the habitats in which they are found. Africa's particular patterns of biodiversity are to a great degree products of its geological history that created its microclimates and topography. About three-quarters of Africa's species of plants and animals have wide ranges across the continent's deserts, savanna, forests, and uplands. Another one-fourth, however, are localized species that have evolved within distinctive environmental enclaves or "centers of endemism".[4] Places with sharp gradations of elevation or contrasting wet and dry conditions such as the Ethiopian highlands and the Cape of Good Hope created biological enclaves or ecologically distinct "islands" where local species, and subspecies could develop. The Cape hosted a particularly large number of endemic plant and bird species while the Ethiopian highlands with longer and denser human settlements generated important endemic cereal grains and oilseeds. Equally, areas like the western sahel are rich in varieties of species even though the total size of the plant and animal populations does not rival that of areas with more available moisture. Chapter 6 treats the biodiversity implications of changes in Ghana's forest/savanna frontier.

In most categories of endemism Africa ranks well below other world areas such as the Amazon basin and southeast Asia in plant and higher vertebrate species, though popular concerns about the cheetah, mountain gorilla, and black rhinoceros have achieved a wide notoriety. Among world areas Africa is richest in the biodiversity of freshwater lakes, with species of cichlids being over 90 percent endemic in Lake Malawi and Lake Tanganyika. Within Africa, the areas of highest density for biodiversity in flora and fauna are the rain forests of West Africa and Zaire. Recent rapid changes in the climate and land use in Africa have resulted in threats to specific parts of Africa's biodiversity, most dramatically in lake fish species where eutrofication and anoxia (lack of oxygen and increased nutrient loads) have sharply reduced species numbers.[5]

In terms of Africa's environmental history, however, endemism may be a less important issue than Africa's absorption and adaptation of genetic stock of domestic animals and food plant germ plasm from international ecological networks. Africa's exchanges of mammal groups began 30 million years ago when the continent temporarily touched the Eurasian land mass and anthropoid primates (e.g., early humans) and elephants migrated out and ancestors of bovids, horses, rodents, and carnivores made their first entrance into Africa. In more recent times, about 7000 years ago, other migrants to Africa from the northeast arrived via the Nile Valley. These were domesticates of the wild ox, *Bos primigenious*, which mixed with local wild variants and composed Africa's first domestic cattle, *Bos taurus*, a humpless longhorn variety, which ranged into the western part of the continent and evolved into several breeds known in Africa today, especially the N'Dama, Kuri, and West African shorthorn. The N'Dama in particular over time developed resistance to local diseases, including the deadly trypanosomiasis (sleeping sickness). Eastern and southern Africa's bovine ancestors, the humped *Bos indicus*, or zebu, arrived first around 1500 B.C. and then in large numbers after the seventh century A.D. The zebu and their descendent varieties developed a tolerance for arid conditions that allowed their spread along the entire Sahelian zone and through eastern and southern Africa. In the third century A.D. another domesticated animal, the camel, made its appearance, introduced by Romans to North Africa. African peoples—the Tuareg, Somalis, Beni Amer, and Nubians—in the Sahel desert zone, the Horn of Africa, and the Nile Valley quickly mastered camel husbandry and adapted their economic culture around the animals' food and labor potential.

Imported domesticated plants have also increasingly provided the bulk of Africa's vegetative cover. Wheat and barley arrived early in the northeast to complement endemic food crops there. Bananas, imports from southeast Asia, arrived via Indian Ocean trade networks, likely in the first millennium A.D., and quickly adjusted to the ecology of the lakes region as a staple food. Human hands induced their spread from Buganda on the northeast corner of Lake Victoria to most areas around the northern edge of the lake by the twentieth century. Propagated by cuttings rather than seed, banana genetic stocks diversified rapidly and adjusted to drier conditions and new soils on the western side of the lake around the Kagera river.[6]

Of all Africa's food plant imports the most pervasive and influential have been the introductions from the Atlantic world: manioc (cassava) and maize. These two food crops in particular, one a forest tuber and the other a variety of grass, have spread throughout the continent since their introduction around 1500, maize becoming the staple food of most of southern and eastern Africa by the second half of the twentieth century. Cassava replaced local yams and sorghum in most of the humid zones of West and Central Africa and continues its migration into new zones in the twentieth century.

Imported nonfood crops have also spread quickly over Africa's land surface. Exotic tree species such as eucalyptus, white pine, and black wattle now cover much of southern Africa as plantation crops or as invasive volunteers. Eucalyptus is the most common tree on the Ethiopian highlands since its introduction only a century ago. Its blue-green leaves have become emblematic of human settlement across the highlands. In eastern and southern Africa imported cash crops such as sisal, pyrethrum, and sugar cane occupy vast monocrop plantations. Cocoa, of Brazilian origin, is a key cash crop in several economies of West Africa. African native crops like cotton and coffee have re-entered the continent in varieties developed externally. Rubber trees, imports from southeast Asia, became widespread in Liberia and Zaire and played a significant role in their modern history.

PATTERNS OF CLIMATE, PATTERNS OF HISTORY

Beyond its geomorphology and its genetic diversity in vegetation and animals, Africa's landscapes reflect its changing patterns of climate. Unlike temperate zones in which growing seasons and cycles of life respond most directly to fluctuations in temperature, Africa's rhythms of life reflect primarily the availability of moisture, especially rainfall.[7] Africa's annual patterns of rainy and dry seasons, the length of each year's growing season, humidity, and soil moisture result from the annual rhythms of cyclonic winds, ocean temperatures, and the earth's rotation around the sun. Following the tilting of half the earth's latitudes toward the sun in summer and away in winter, the anticyclonic and trade winds set the yearly cycle between rainy and dry seasons. This shifting zone of rain-bearing turbulence, which climatologists call the Inter-Tropical Convergence Zone (ITCZ), sets a general two-part (bimodal) pattern of seasons, one wet and one dry, which characterizes the continent as a whole. This seasonality also, however, produces subtle variations from year to year and in particular places according to elevations, topography, and global climate anomalies such as El Niño (or ENSO) and tropical Atlantic surface circulation.[8] Several years of short or delayed rains constitute drought, historically a common occurrence in much of Africa.

Africa's annual cycle of weather is seasonally predictable, if at times erratic. The winds that bring moisture to Africa north of the equator move in patterns from south to the north of the equator with the rotation of the earth toward the sun in the summer months (June to September). The movement of the turbulence north of the equator brings summer rains in the northern hemisphere. The onset of the rains has a remarkable effect. Within two weeks brown, lifeless landscapes turn green, seeds germinate, and chemical reactions within soils make nutrients available to plants. From December through March air masses from the north dominate, creating a long dry season as the

Figure 2.2   Blue Nile Falls (Tisissat), September 1973 (rainy season).

rain-producing ITCZ turbulence moves south of the equator. In the dry season fields ripen for harvest, pasture grasses shift into dormancy, and livestock migrate to pasture near water sources. The Serengeti's dramatic movement of wildebeest is a well documented example of this annual effect. Locally, the ITCZ's movement interacts with local topography to result in moist slopes or rainfall shadow effects that have effects locally in crop choices, land values, and landscape textures. At opposite ends of the continent at the Cape and along the northern litoral a Mediterranean climate prevails, making wine, citrus, and grain production possible.

   Vegetation patterns, animal movements, and human economies in Africa have adjusted to these repetitive patterns over several millennia. Over the longer geological time frame Africa's climate has been more often dry than wet and more often warmer than cooler. In that time scale, expanding desert zones may have connected the Sahara to the Kalahari along arid corridors through a shrinking Central Africa forest. Many arboreal African species such as hyraxes, squirrels, and monkeys actually descended from ancestors forced to survive on the ground as tree cover diminished.[9] Within the last few millennia, however, there have been wetter periods, including one in which the Sa-

Figure 2.3   Blue Nile Falls (Tisissat), May 1974 (dry season).

hara was a pastoral grassland that supported cattle, game, and human settlement. This period ended somewhere in the middle of the third millennium B.C. (c. 2300 B.C.).

These long-term fluctuations have been significant during epochs of human history in which African social institutions and economic strategies evolved with the experience of human communities and individual farmers, hunters, and pastoralists. Iliffe generalizes that Africa's agricultural systems were historically mobile, a strategy to adapt to the environment rather than transform it. Yet, as always, there has been an infinite set of variations and exceptions. In northern Ethiopia farmers did not move with the seasons but rotated the planting of cereals and pulses (peas and beans) in a fashion finely tuned to the arrival and departure of summer rains followed by a long dry season. In eastern Sudan, camel-herding Hadendowa planted sorghum in moist riverain soils and then retreated with their animals to distant wet season pastures until the harvest was ready eight months later. Farmers in northeast Zambia practiced *citamene*, a system of shifting agriculture that required large tracts of land but maximized scarce soil nutrients. In East Africa the Maasai historically followed pasture with their herds of cattle and goats, carefully avoiding full contact with

thickets of acacia scrub infested with tsetse flies bearing deadly sleeping sickness (trypanosomiasis).

As Iliffe suggests, most livestock economies in Africa chose to follow the movement of moisture and pasture rather than stockpile their
animals' food supply. Choices to migrate by herders of camels, cattle,
sheep, or goats were finely tuned judgments about those species' needs
for food and moisture, as well as their capacity to work and serve as
a form of wealth. Not only did Africans wait for nature to alter their
physical setting, but in most areas farmers and pastoralists used fire
as a strategy to clear fields, raise phosphorus and Ph levels in the soil,
control disease, and stimulate the growth of new pasture grasses. Over
the course of Africa's landscape history, fire was the primary tool that
shaped vegetative cover either through conscious human planning or
the serendipity of lightning strikes.[10]

The seasonality of moisture affected not only food supply but also
disease, migration, the timing of military campaigns, and ritual cycles
in politics and religion. Most tropical diseases, whether epidemic or
endemic, were seasonal events. Military states generally organized
raids and expansionary campaigns during the dry season when their
soldiers were free from agricultural work and enemy harvests were in
storage. For many African societies the months just prior to harvest
were the "hungry season" devoted to preserving energy in both humans and livestock for the harvest season ahead.

Laid against the annual cycles of weather, however, are the longer-
term shifts in climate and ecological conditions. In recent years historians and climatologists have tried to assemble evidence of climatic
epochs in African history that may shed light on the historical economic trends and political events. Sharon Nicholson, an historical me-
teorologist, was among the first to reconstruct the climate history of
Africa from historical sources. Using historical accounts, evidence of
lake levels, and indicators of past climate conditions, she argued that
the West African Sahel experienced several climate epochs over the
course of the A.D. 800–1600 period when, in succession, the Sahelian
empire states of Ghana, Mali, and Songhay developed, thrived, and
declined. She hypothesized that the period c. A.D. 800 to 1300 was
relatively wet, followed by the drier 1300 to 1450 span and followed
again by a wetter period from the late fifteenth to the late eighteenth
century.[11] The question of using such evidence to assess the environment's role in African history differs substantially from the kind of
environmental history implicit in Iliffe's thesis of African history as

the human struggle against nature. To what extent did environment *determine* the events and patterns of human history in Africa? Or did environmental factors simply fix the dimensions of the playing field and help set the rules of the game? Chapter 3 examines examples of major events and features of African history by which it may be possible to assess the environment's role in shaping historical change.

DEMOGRAPHY: HUMAN HANDS

Africa's population, growth, density, and rate of fertility is an essential element of the continent's environmental history, particularly from the perspective of anthropogenic landscapes (see Chapter 1). Humans' historical ability to effect changes in soils, vegetation, and biodiversity is fundamentally a question of the concentration of labor at particular points of time and space and the nature of technology— the combination of tools, knowledge, and experience. Iron tools, either produced locally or obtained in trade, for example, were in wide use in Africa during the period treated in this book. Iron-tipped plows that raised labor efficiency in agriculture had long been in use in Ethiopia and Eritrea but only appeared in southern and western Africa in the nineteenth and twentieth centuries and magnified the human impact on the land. In most areas of Africa, however, preindustrial economies depended on the direct imprint of human labor for economic activity, and the distribution of labor on land has been the most significant feature of humans' effects on landscapes.

It is beyond doubt that Africa's population has grown overall through time, but it is the question of the population's rate of growth, decline, age structure, morbidity, and fertility rates at particular points in history that matters to environmental history. Debates over historical demography—and especially fertility rates—in Africa has produced much more heat than light since the dearth of hard historical data has given an exaggerated weight to theory as the tool of argument. For Africa the dominant "natalist" perspective, based largely on demographic transition theory derived from Western Europe's historical data, has argued for relatively slow or stagnant population growth historically because high fertility rates have matched mortality from disease, poor nutrition, and political violence. This theory asserts that Africa's recent rates of growth between 3 and 4 percent has resulted from the benefits of colonial hegemony: improved health care, famine relief, and, at least until the post–Cold War period, imposed peace. The natalist view also holds that at some later date fertility rates and population growth will decline in response to economic development as they have in Western Europe and North America.[12]

The opposing "antinatalist" argument derives from the more empirically based evidence of preindustrial England, where birth records and death rates indicate that population growth resulted from social responses to economic conditions, that is, rural families exercised control over fertility through social actions such as late marriage and celibacy. Antinatalists thus begin from the premise that fertility is a function of social control and not biology. In their view, then, Africa's historical growth rates have been specific responses to conditions like colonial rule or economic change and not simply to changes in death rates. In John Thornton's work on eighteenth-century Kongo baptismal records, where there are birth and death data comparable to Western Europe, the evidence shows slow growth tempered by occasional shocks of disease or temporary food shortages.[13]

While demography is an essential element of any environmental history, we must acknowledge two facts. First, the direct historical data do not exist for the continent as a whole, though there may be rare exceptions, like the Kongo baptismal records. Second, general data on Africa as a whole has little value for understanding specific local and regional contexts. The issue for human impact on an African historical landscape is the number of people at a place at a certain time—population conjuncture—and not a simple growth rate. After all, people move. They migrate to or leave an urban area and congregate on productive land, or are coerced to do so, at a rate much higher and more significant than an overall population growth rate. The historian's task then is to detect evidence of movement and density of human population. The effect of population is therefore situational and fluid and not amenable to generalization.

John Illife's environmental thesis gives a high priority to the harsh rules of the game set by Africa's physical setting. Climate, topography, soil, and human population movements indeed provide boundaries. Yet, the possibilities of conjuncture and human economic culture are nonetheless virtually infinite. Chapter 3 examines the extent to which these boundaries have defined and shaped particular African historical events.

NOTES

1. John Iliffe, *Africa: History of a Continent* (Cambridge, 1995), 1.

2. Jonathan Kingdon, *Island Africa: The Evolution of Africa's Rare Animals and Plants* (Princeton, 1989), 146–48.

3. J. Peter Sutcliffe, "Soil Conservation and Land Tenure in Highland Ethiopia," *Ethiopian Journal of Development Research* 17, no. 1 (1995): 63–87.

4. Endemism is the process of formation of distinctive species of flora and fauna that takes place in specialized habitats. See Kingdon, *Island Africa*, 4.

5. On lake endemism see, Les Kaufman, "Catastrophic Change in Species-Rich Freshwater Ecosystems," *BioScience* 42, no. 11 (1992): 847; on biodiversity in general, see Brian Goombridge, ed., *Global Biodiversity: Status of the Earth's Living Resources* (London, 1992), 129–42.

6. C. C. Wrigley, "Bananas in Buganda," *Azania* 24 (1989), 64–70; Gerda Rossel, "Musa and Ensete in Africa: Taxonomy, Nomenclature, and Use," *Azania* 29–30 (1994–95), 130–46; E. DeLanghe, R. Swennen, and D. Voylsteke, "Plantain in the Early Bantu World," *Azania* 29–30 (1994–95): 147–60.

7. Eugene Rasmussan, "Global Climatic Change and Variablility: Effects of Drought and Desertification in Africa," in *Drought and Hunger in Africa: Denying Famine a Future*, ed. Michael Glantz, (Cambridge, 1987), 3–22.

8. See, for example, Peter J. Lamb, "Large Scale Tropical Atlantic Surface Circulation Patterns Associated with Sub-Saharan Weather Anomalies," *Tellus* 39 (1978): 240–51.

9. Kingdon, *Island Africa*, 15.

10. The best overall work on the role of fire in human landscapes is by Stephan J. Pyne, *Vestal Fire: An Environmental History, Told Through Fire, of Europe and Europe's Encounter with the World* (Seattle, 1997). No comprehensive work yet exists for Africa, but see Fairhead and Leach, *Misreading the African Landscape: Society and Ecology in Forest-Savannah Mosaic* (Cambridge, 1996).

11. Sharon Nicholson, "A Climatic Chronology for Africa," 75–81, 251–54, cited in James Webb, *Desert Frontier: Ecological and Economic Change along the Western Sahel, 1600–1850* (Madison, 1995), 4–5.

12. For the best summary of the debate, see John Iliffe, "The Origins of African Population Growth," *Journal of African History* 30 (1989): 165–69.

13. A fundamental debate still continues between historians who attempt to assess the effects of the Atlantic slave trade on African population. See, for example, C. C. Wrigley, "Population in African History," *Journal of*

African History 20 (1979): 129–31, and Iliffe, "Origins," 167–68. See also John Thornton, *Africa and Africans: The Making of the Atlantic World,* (Cambridge, 1996), 72–73, and Patrick Manning, *Slavery and African Life: Occidental, Oriental, and African Slave Trades* (Cambridge, 1990), 38–85.

Chapter 3

Environment and History in Africa

Africa's history is first and foremost the history of its people, their interactions with one another and the land. The social and political context of that interaction is as varied as the social life of Ibo villages, the seasonal migration of Somali camel herders, and the court life of a Bugandan *Kabaka*. The rise and decline of large, powerful state systems has nevertheless been a consistent theme in writing about African history. The bulk of that coverage, however, has focused narrowly on the stage of politics and exchange (especially international trade). This chapter poses the question of whether an environmental lens might offer a new and enriching perspective on Africa's impressive political histories.

This chapter takes several celebrated cases of state building in Africa as opportunities to analyze the implications of environmental factors in explaining the rise, flowering, and decline of historical formations in Africa. The cases are the Sahelian states, especially the Mali empire, Great Zimbabwe, and Aksum. Each of these cases represents a particular historical epoch, a distinctive regional setting, and a body of evidence about Africa's history.

TRADE AND STATE BUILDING IN THE SAHEL

Between the fifth and the seventeenth century A.D. several large state systems appeared in the areas between the Sahara desert and West

Africa's forest belt. The most celebrated of these state systems—Ghana (c. 400–1250), Mali (c. 1250–1492), and Songhay (c. 1492–1591)—were multiethnic empire states that developed large, hierarchical political cultures and regional economies that have come popularly to symbolize the glories of the African past and its capacity to match or even surpass the scale of Eurasian empire states. Mali and Songhay in particular witnessed an efflorescence of oral and written expression that left deep impressions on the historical record locally, in the Mediterranean, and within the wider Islamic world. Though each of the states had its own distinctive geography, central historical charter (Mali's Sundiata epic being the most celebrated), and military tradition, they nonetheless shared features of their core economies and political structures as central state systems based on tribute paid in kind from smallholder cereal farms and on revenues skimmed from the movement of high value to weight goods traded across the Sahara and into the Mediterranean regional economy. Chief among these goods were gold, slaves, cloth, and leather. These states also shared an ecological world that stretched between the desert's edge and beyond the cattle zone of the Sahel and the savanna, reaching south to the woodland mosaic landscapes that bordered West Africa's rain forests. Within the historical record we have a fairly deep appreciation of their political culture and images of their physical setting but little real grasp of the fundamental factors of economy and ecology that fostered their growth, daily rhythms of life, and decline. To what extent did the rise, decline, and evolution of political and economic power of these far-flung empires depend upon conjunctures of environmental change? Was the environment a cause or merely a context for historical events?

Beginning from Sharon Nicholson's climate data on Africa, two historians, George Brooks and James Webb, have extrapolated and defined two contrasting patterns of environmental and historical change in the West African Sahel. Both historians delimit key West African historical zones not by fixed geographical boundaries but by shifting rainfall boundaries (isohytes) that generate particular types of vegetation and impose specific constraints on human activity. The 100-millimeter rainfall zone separates the Sahara from the Sahel, while the 400-millimeter line demarcates the Sahel's southern edge and the zone where cultivation of drought-tolerant crops like sorghum and millet is possible (see map). South of the 400-millimeter line is the savanna itself, the zone that Mande oral narratives call the "bright"

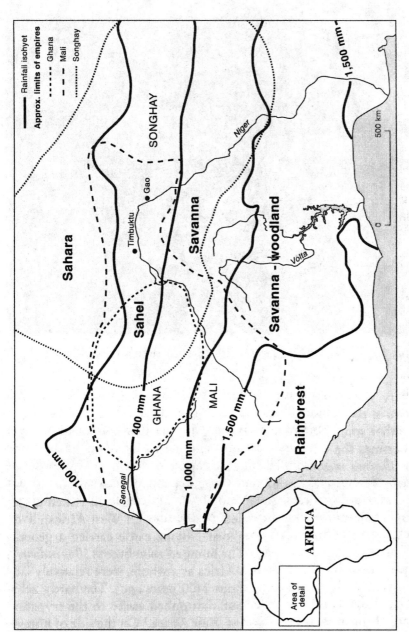

Figure 3.1 Map of West African Empires with Rainfall Isohyets

Figure 3.2   Ndama (trypanosomiasis-resistant) cattle, Kumasi (Ghana), 1997.

country where Sundiata and his royal Keita clan founded the Mali empire in the middle of the thirteenth century.

Farther south, below the savanna, lies the 1000-millimeter isohyet that defines the woodland savanna and marks the limits of the tsetse fly (*Glossina morsitans*) habitat, a barrier over which livestock (especially horses and cattle) and those dependent on them dared not trespass. Cattle, horses, sheep, donkeys, and camels all risked death from the sleeping sickness carried by the flies. Of West African livestock, only the humpless longhorn Ndama cattle carried a genetic resistance to trypanosomiasis. The humped zebu breeds (*Bos indicus*), which are the most common in Africa as a whole, were relatively late arrivals in West Africa (only about 1400 years ago). The hardy zebu tolerated dry conditions well but succumbed easily to the trypanosomes of the more humid areas of West Africa.[1] On the side of human disease, malaria's domain also moved with the shifting seasonal bands of moisture and over the longer-term, movements of rainfall boundaries. Food supplies in their type and reliability also historically reflect moisture. Sorghum and millet, West Africa's historical cereal crops, are drought resistant but long maturing and offer relatively low yields per unit of land. Maize, which arrived after 1500, offered much higher yields and was ready to harvest or consume green months earlier. Maize flourished in many areas but quickly disappeared in other areas in the nineteenth century when annual rainfall fell below 400 to 500 millimeters.[2]

What George Brooks's work adds to our understanding of the history of the Sahel and its great empire systems is his connection of specific human activities, such as cavalry raids and iron smithing, to the dynamic of shifting and swaying ecological boundaries across the West African historical landscape. During periods of less aridity the agricultural frontier moved north, allowing cereal-producing farmers to expand their settlements into the Sahel, thus pushing the political frontiers of their ruling elite farther north toward key trans-Saharan trade routes. In these times of greater moisture the domain of the tsetse also crept north, restraining the mobility of horse-borne troops who attempted to raid peoples at the southern edge of the savanna. The 1000-millimeter isohyet, in effect, provided a protective screen for the agricultural peoples who had often fallen victims to slave-raiding horsed warriors in drier conditions. In fact, tsetse provided an effective defense against invasion. Brooks points out that over the past two millennia, the tsetse zone has fluctuated as much as 200 kilometers north to south, thus altering the terms of livestock economy and military advantage. Much of the Mali empire's historical expansion south closer to sources of gold and kola (e.g., between A.D. 1250 and 1450) took place during a generally dry period that allowed the cavalry to push Mali's formal hegemony to the edge of the forest whence their merchants influenced newly emerging forest kingdoms in the eighteenth century (see map above).[3]

Mali's economic and intellectual centers at Jenne and Timbuktu emerged in their regional roles during what Nicholson describes as a generally humid period when agricultural and trade frontiers may have advanced north and emphasized the key positions of these urban centers. Brooks, however, sees the environment's effects differently. He argues that the 1000 to 1500 period was dry, encouraging Mande smiths and merchants south into the savanna woodlands where they planted the seeds of the Mali empire's enduring cultural diaspora. The retreat of the tsetse barrier as much as 200 kilometers south, according

Figure 3.3   Timbuktu (a nineteenth-century view). From Henry Barth, *Travels and Discoveries in North and Central Africa, 1849–1855*. New York: Harper and Brothers Publishers, 1859. Vol. 3.

to Brooks, enhanced Mali's military interests, allowing the empire's military to use their cavalry up to the edge of the forest. Brooks thus sees the decline of Mali and the rise of Songhay in similar environmental terms. Dry conditions in the Sahel forced Tuareg camel pastoralists south where they established control of Timbuktu in 1483, and the southern movement of the tsetse zone allowed Mali's southern enemies, such as the Mossi, to mobilize horsed warriors against Mali. In this view Songhay succeeded Mali because of its control of the Niger River and its horsed military. Finally, Brooks argues, the brief wet period from 1500 to 1630 set the terms for both Songhay's enrichment through trade and its demise since wet conditions along the caravan route expanded trade networks across the desert but eventually tempted the expansion-minded Moroccan king al-Mansur to launch a conquering army across the desert to capture what he thought was the golden goose of the trans-Saharan trade. Brooks sees quite specific effects:

> The wet period brought a mixture of benefits and disadvantages for the Songhai empire. Trans-Saharan commerce flourished during this time

and the Songhai empire was able to prevail over Berber groups and control trade routes further north than either Ghana or Mali had been able to penetrate. Ultimately, the favorable conditions along the trans-Saharan routes allowed Moroccan forces in 1590–91 to conquer Songhai. Moroccan forces could not penetrate further south because of the disease affecting their cavalry. The wet period marked an era of transition during which the commerce of each of the climate zones of western Africa was increasingly reoriented from the interior to the coast, generally to the disadvantage of the Mande speaking groups.[4]

This neatly packaged view of the environment's effect on history is appealing since it links events directly to environmental factors. Yet, without clear and more precise data on the period 800–1600, tying historical events and trends to climate is a misleading exercise. Brooks's data offer plausible historical arguments about the effects of the environment on historical events, but they fundamentally disagree with Nicholson's evidence about the sequencing of "wet" and "dry" epochs. Nor does Brooks's environmental determinism account for the year-to-year and season-to-season variation in rainfall that recent farming systems research indicates profoundly affects both agriculture and pastoralism.

The real value of linking environment to historical process may lie in a more subtle, nuanced view of how environmental conditions set a context for social and historical interaction. Sharon Nicholson and George Brooks argue for a generalized pattern of climate, that is, that between A.D. 1100 and the mid–nineteenth century there was a general trend toward more aridity with the exception of a century and a half of more humid conditions from c. 1500–1630. James Webb, a historian of West Africa's arid zones, adds subtlety to the role of ecology in the formation of human history in West Africa. From historical records for the period after 1600, however, Webb persuasively documents a consistent trend toward a drying climate.[5] Rather than attempt to identify specific events of human history related to climate, however, Webb describes the dynamics of agro-ecological zones such as the "Sahelian Cattle Zone" and the "Great Camel Zone" whose movement he traces over time and whose dynamics he describes in terms of the ecological limits of the animals at the core of the human livestock economy. His approach emphasizes the fluidity of movement over time of these ecological economies that underlay Sahelian empires and human actors. His data, for example, give specific evidence of how the camel's virtues changed human capacity for life in a drier environment:

> The camel could survive without water or fresh grazing for 8–10 days, or twice as long as a donkey; the strongest draft camels could carry 200 kilograms of freight, or one-third more than a donkey or the draft ox; fewer men were required to tend camels than donkeys, and thus long-distance transport costs, figured in ton-kilometers, were lower. . . . The milch camel's lactation was not dependent upon the availability of fresh pasture, as was the lactation of cattle, sheep, and goats. Milch camels normally lactated for 11 months out of 12 and the camel's ability to turn salty water into sweet mik for almost the entire year allowed desert people to exploit lands which otherwise would have remained sub-marginal.[6]

Webb's reconstruction of climate patterns allows him to argue that in the year 1600 the major zones of camel herding, cattle herding, and rain-fed agriculture were approximately 200 to 300 kilometers north of their position 250 years later.

By 1850 the Sahelian Cattle Zone had descended south into lands that had once supported agriculture; the farming frontier had retreated before it. In other words, pastures at the northern edge of the Sahel that had nourished cattle at the end of the Songhay empire supported only camel herding by the opening of the colonial period.[7] Such changes also have interesting and important implications for human geography, seasonal livestock movement, and the calendar of human and livestock disease. What the first generations of European colonial forces saw when they arrived in West Africa was therefore not age-old structures but a fairly new configuration of human settlement. Key historical events like the nineteenth-century jihads of Usuman don Fodio and al-Hajj Umar therefore took place in a setting of eco-logical adjustments that, if they did not determine the movements' intellectual foundations, at least must have influenced military field strategies and the logistics of feeding the faithful.[8]

Webb points to the relationship, not of specific historical events and ecology, but of environmental change and the cultural processes that dominated the savanna/Sahel zone in the twilight years and after the decline of the great empire states. He argues that the general shifts in climate/ecology/economy that took place between the fourteenth and the seventeenth centuries brought desert, savanna, and Sahelian peoples in closer contact and produced a new historical identity that blended the economies and historical traditions of the Arab, Amazigh (or Berber), and Mande peoples. The growth of Timbuktu from a community of 10,000 persons in 1325 to a commercial/intellectual center of 30,000 to 50,000 people in 1591 was part of this process rooted in ecological change, but not directly attributable to it.

Yet, Webb also describes a darker side of the emerging regional economy of the seventeenth century. The trade and political relations that emerged between the peoples of the southward-moving desert's edge and the peoples of the savanna had harsh consequences for many as a regional slave and horse trade expanded in the seventeenth century—a trade that, for a time, rivaled the trans-Atlantic trade from the West African coast. The means of production was military strength and mobility (wherein the camel played a large part) and produced human booty. As Webb states: "Viewed over the long term, it appears that, as the southern shore of the desert moved southward and waves of political violence broke onto the savanna, Black Africans were caught in the undertow."[9] Ecological change in this case expanded zones of conflict and the loss of human freedom.

Taken from the perspective of environmental history as a whole, however, the political history of the Sahel assumes a new richness. The physical environment of the great empire states was not a fixed canvas, but a shape-shifting stage that demanded a continuing set of adaptations of economic base and political structure. If anything, an appreciation of the environmental context of the history of the West African savanna enriches our appreciation of the skill of leadership and the social dynamism of West African political culture. When the states of the Niger valley finally succumbed at the end of the sixteenth century, the end was due to a wider set of factors than an invading army from Morocco or even an impinging desert edge. These factors included, *inter alia*, decreasing farm productivity, the retreat of livestock herds in favor of camel husbandry, and the loss of population to slavery. If the new forms of politics and economics pointed out by Webb's work were a grim epilogue to an historical process of building, they were also a prologue to the birth of the new Atlantic world in which Africa and a new ecological stage played a major role.

GREAT ZIMBABWE: A CHALLENGE IN SOUTHERN AFRICA

Great Zimbabwe is the name commonly given to a collection of more than fifty ruined settlement sites on the Zimbabwe plateau at about 1000 meters above sea level. The remains of these settlements contain outstanding examples of mortarless granite wall construction as well as silent testimony to the existence of both a concentrated urban setting and an expansive state structure that had its greatest expression in the period A.D. 1250 to 1450, that is, contemporaneous with the Mali empire.[10] Unlike Mali, however, the state of Great Zimbabwe had boundaries somewhat fixed by its topography and left the physical remains of its concentrated wealth and elaborate material culture in the form of impressive stone structures. The site itself had a clear geography: bounded in the north by the Zambezi River valley, in the south by the Limpopo river, and in the west by the Kalahari, in the east the escarpment of its plateau opens onto the coastal plain leading to the coastline of Mozambique and the Indian Ocean.

While Mali's trade network linked the West African Sahel to the Mediterranean world, Great Zimbabwe merged its location above the Zambezi valley to an Indian Ocean trade network that, in the fourteenth century, had prompted Swahili city-states to sprout mushroomlike along the East African coast in places like Kilwa, Lamu, Mombasa, and Malindi. Great Zimbabwe's coastal entrepôt was probably the almost mythical port of Sofala on the Mozambiquan coast, a destination sought for many years by the Portuguese as an African El Dorado. While no written record exists from a contemporary visitor to Great Zimbabwe, Ibn Batutta, the great fourteenth-century North African traveler who also described the Mali empire, stated that Sofala's source of gold was a site a month's march inland from the coast; doubtless he was describing Great Zimbabwe's gold workings.[11] Nevertheless, Portuguese accounts describe only the waning years of Great Zimbabwe and offer no eyewitness accounts. The Shona people of modern Zimbabwe, the descendants of the Great Zimbabweans, retain some oral traditions but none of the fixed texts of bards or Arabic records of travel such as exist for empires of the savanna.

Though archaeology offers numerous shreds of evidence on the origins of Great Zimbabwe, the best evidence is, in fact, ecological and circumstantial. First, the evidence of settlement on the Zimbab-

Figure 3.4   Great Zimbabwe royal enclosure.

wean plateau and the construction of stone structures coincides with similar dates for the rise of the Indian Ocean trade system in East Africa, that is, as early as the ninth century, and the growth of coastal city-states in the second quarter of the present millennium. That trade system linked East Africa to trade networks from the Red Sea, Persian Gulf, the Indian subcontinent, and China. An excavation at "Renders Ruin" at Great Zimbabwe itself (an area thought to have been occupied by the wives of the king) unearthed a hoard of imported luxury goods: a glazed Persian bowl, fragments of engraved Near Eastern glass, bronze bells, cowry shells and several thousand glass beads. At another spot nearby archaeologists found a coin minted at Kilwa, the East African coastal trading city.

Economic incentives to trade in luxury goods and metals nevertheless depended on the ecology of travel. Trade contacts with the coast moved overland, likely along river valleys, to connect with coastal traffic on wooden ships: lateen-sail–powered dhows that moved up and down the coast with the same monsoons that regulated Africa's seasonal patterns of rainfall.[12] East Africa's primary products offered to the trade were diverse, but only Great Zimbabwe could have contributed gold in any appreciable amount. One historian has estimated that between seven and nine million ounces of gold circulated through the East African trade in the period before the arrival of European colonists on the plateau in 1890. Based on the evidence available today, Zimbabwe would have been the only large-scale source of that metal.[13]

As was the case in the West African savanna, external trade built upon a local capacity to sustain supplies of labor, food, and a political/religious elite. More circumstantial evidence points to the ecological context of state formation. Like Mali and Songhay, Great Zimbabwe occupied a zone that allowed it to withstand a harsh seasonality dominated by the movements of the Inter-Tropical Convergence Zone into the southern hemisphere. The Zimbabwe plateau has a wet season from about November to March, followed by a dry season April to October. Except for its elevation, the savanna woodland vegetation is similar to the southern reaches of the West African savanna. Rainfall was sufficient to support indigenous crops like sorghum, millet, and squashes but was fickle: the historian David Beach estimates that one season in five suffered from drought or other sources of crop failure.[14]

How did Great Zimbabwe concentrate sufficient supplies of food to maintain a labor force of masons, gold miners, farmers, and a polygynous royal household? The archaeological record has a clear answer: cattle, in apparently large numbers. On the lower slopes of the hill at Great Zimbabwe, archaeologists found a midden containing about 140,000 pieces of bone. Of the over 15,000 bones analyzed, all but 218 were cattle bones! Moreover, the age of the cattle (fairly immature) suggests that these were herds kept close to home and not pastoral herds that migrated seasonally. Great Zimbabwe's royalty obviously enjoyed young and tender beef.[15]

Indeed, a major characteristic of the plateau that distinguishes it from the surrounding areas of Central Africa is that it is tsetse free and surrounded by lowland grazing zones that are seasonally free from the deadly fly. These cattle thus likely provided the meat and milk, as well as a means of stored wealth to sustain a concentrated population through a crop failure or shortfall in seasonal rains. We may also assume that cattle served, as they have historically in world history, as social capital used for tribute payments, bridewealth, and symbols of elite consumption. In the pattern common in African history, manipulation of cattle resources allowed the concentration of power among elites, elder over junior, and male over female. Shifting terms of trade between grain and cattle also provided a means of stabilizing food supplies during drought or other kinds of harvest short-falls.

The archeological record for Great Zimbabwe's decline is less suggestive than for its success. By 1500 the Zimbabwe state, like its trade partners on the Indian Ocean coast, had largely disappeared from the scene with few traces beyond its silent dry stone monuments. The shallow gold mines appear to have been abandoned or simply played out. In the absence of other evidence some archaeologists have concluded that Great Zimbabwe fell victim to its own success: the large concentration of population on the plateau overextended its resource base. Graham Connah summarizes this view:

> The apparent reasons for the decline and abandonment of Great Zimbabwe and its associated sites consist of a reversal of the factors that gave rise to their growth. The gold trade declined, probably because of falling world prices and the depletion of the more easily worked deposits on the Zimbabwe Plateau. More important, however, the environment around Great Zimbabwe collapsed: overcropped, overgrazed, overhunted, overexploited in every essential aspect of subsistence agriculture, it ceased to be able to carry the very concentration of people that it had given rise to.[16]

In the absence of hard evidence, Connah's Malthusian hypothesis may seem as plausible as any other. Yet, his environmental hypothesis for Great Zimbabwe's precipitous decline emphasizes human perfidy (overgrazing, overhunting, etc.) and assumes an inability to adapt to population pressure. The decline of world gold prices also seems improbable given that Great Zimbabwe was already a ruin by the end of the fifteenth century, that is, prior to the sudden arrival of New World gold on the world economy.

I would argue for a different set of possible environmental causes for the decline of a state based on cattle, cereals, and the local production and long-distance trading of gold. One hypothesis can take as evidence archaeological findings on the over 1200 mine workings in the area. The archaeologist Tim Huffman has argued that the dates of gold mining on the plateau coincides almost exactly with the dates of Great Zimbabwe. The presence of shallow quartz reef mines, none more that 25 meters deep, suggests not only that gold was a major source of wealth, but that there were limits—problems of flooding and ventilation—that seem to have prevented further development of these mineral resources.[17]

But Great Zimbabwe's demise might as easily have come in a crisis of local production. The temperature and rainfall conditions that made settlement on the Zimbabwe plateau ideal may not have been stable over the longer term. The delicate balance of the agro-pastoral economy which sustained social reproduction and local food supplies would have been vulnerable to the same shifting climate conditions that took place in the Sahel in the same period, that is, increasing moisture from about 1500 and/or the serious droughts evident in the region in the late twentieth century.

Great Zimbabwe's cattle-based social economy would have been particularly vulnerable to an advancing tsetse frontier since the archaeological evidence indicates that Great Zimbabwe's cattle were Sanga, zebu cross-breds notoriously susceptible to trypanosomiasis. Unlike later Maasai zebu herds in East Africa that historians like Richard Waller and James Giblin argue had controlled contact with the disease sufficient to develop immunity, the isolation of Zimbabwe cattle from contact made them particularly vulnerable to any expansion in tsetse frontiers over dry season pasture or onto the plateau itself.[18] Apart from disease, sustained drought such as that experienced in the late 1980s and 1990s in southern Africa might well have broken down the balance of labor, food supplies, and livestock pasture necessary for the high concentration of population on the plateau. Such a high population concentration on the plateau was, after all, a distinct historical conjuncture of economy, ecology, and society in that part of southern Africa, an anomaly rather than the norm. Like other explanations, however, this view remains a hypothesis subject to a generation of new research, especially new evidence from archaeology on the lives of those farmers and herders outside the walls of the royal enclosures.

BLACK SOIL EMPIRE: AKSUM AND THE ETHIOPIAN HIGHLANDS

Between the first century and the seventh centuries A.D. an elaborate, urban-based empire perched on the highlands of Ethiopia and Eritrea, dominating a zone extending from the Red Sea coast in the east to the Nile Valley in the west. Aksum produced outstanding feats

Figure 3.5 Aksum (a nineteenth-century view). From Henry Salt, *Twenty-Four Views in St. Helena, the Cape, India, Ceylon, the Red Sea, Abyssinia, and Egypt* (London, 1809).

of engineering, military prowess, and a sophisticated commercial culture—including its own coinage—that reflected its active participation in the economic world of the Red Sea–Nile Valley–Eastern Mediterranean nexus. Aksum's stone masons and engineers cut, transported, and erected huge stelae of dressed syenite (a fine-grained, grey-green granite) at and around their urban capital. The largest of these stelae, some 33 meters long, may be the largest stone block erected anywhere in the world. In and around Aksum there are hundreds of stelae of various sizes as well as the remains of temples, public buildings, cemetery sites, and burial vaults. One private residence measures 3,000 square meters and came equipped with a bakery and what appears to be a water-drainage system.

These achievements are only the most visible remnants of one of the world's most concentrated and productive political structures. Aksum-related settlements sprinkled throughout the highlands suggest urbanization, a regional network of trade, and a sophisticated management of environmental resources. The archaeological record of the Aksumites' economic, intellectual, and aesthetic world reveals (1) their appreciation for glass beads from Arabia and India, Egyptian glassware, and wine from Roman Gaul; (2) a writing system; and (3) coins that bear the image of their king and, after the fourth century, that include a cross acknowledging their Christian God.

Aksum's prosperity as a state, a society, and economy lay in its geographical setting at the intersection of several trade networks of the ancient world, but also in its own creativity within its distinctive ecological setting. Although Aksum's high urban culture displayed clear evidence of contact with the wider economic world of the Nile Valley and eastern Mediterranean, the Aksumites fed-themselves with home-grown gardens and fields. In fact, we have a much clearer idea about what Aksumites ate—the early highland crop repertoire—than we do about their public life, work, leisure, or life of the mind. The evidence of both language and plant biology indicates that Aksum's home in the Ethiopian highlands was a center of secondary dispersal for a wide variety of crops, especially annual grain crops; Christopher Ehret has used linguistic reconstruction of the Afro-Asiatic language family (which includes Ge'ez, the language of Aksum, and most languages of the highlands) to argue that domestication of teff (a distinctive local grain and eleusine (finger millet) took place on the highlands 7000 years ago.

Aksum's crop endowment was not a royal bequest or an imported benefit, but the hard-won product of farmers' manipulation of a distinctive physical environment. Atop the Shire plateau in what is now Eritrea and the Tigray region of Ethiopia, Aksum had the advantages of an "island ecology" consisting of small ecological niches with descending elevation, subtle gradations of temperature and rainfall, and variations of soil types. In this setting the Ethiopian highlands had become a center of endemism in both vegetation and wildlife. To these natural processes, the ancestors of Aksum's farmers over many generations had selected and propagated specially adapted seed germ plasms and subspecies (land races) of local grasses into a set of distinctive food crops. Thus, the highlands became a primary center for diffusion of several cereal and oil seed crops: finger millet (*Eleusine coracana*), niger seed (*Guizotica abyssinica*), and teff (*Eragrostis teff*). Ethiopia is also especially rich in cultivars of barley, wheat, sorghum, teff and finger millet. N. L. Vavilov's encyclopedic listing of the world's cultivated plants also describes a wide variety of Near Eastern pulses/legumes that are endemic to Ethiopia, suggesting a long period of local cultivation and genetic diversity there. In all, Vavilov names twenty-four grains and oil seeds for which the Aksum area is a world center. Adding to this mix of endemic food crops, at an early stage Ethiopia's highland farmers adopted Near Eastern staple grains such as wheat, barley, and varieties of sorghums from the Nile Valley and elsewhere in Africa. More recently, archaeologists have uncovered evidence that Aksumites also cultivated flax and grapes.[19]

Agriculture is a specialized form of human management of the natural world. By the middle of the first millennium B.C., the Ethiopian highlands had assembled a remarkable arsenal of types and varieties of annual cereals and pulses from which farmers could choose, depending on the elevation, soil type, moisture, and fertility of a particular field. To this witches brew of localized genetic materials, farmers on Ethiopia's highlands added a crucial technology. The technology was the single-tine oxplow, a brilliantly adaptable tool simple in design and light enough to be carried on the farmer's shoulder from field to field (see Figures 3.6 and 3.7). No other sub-Saharan farming system developed the use of the plow for agriculture before the nineteenth century, and nothing has since effectively replaced the local plow on Ethiopia's small farms.

The key complement to the plow was the evolution of a sophisticated tradition of animal husbandry that involved the breeding and training of oxen for animal traction as well as the use of mules, donkeys, horses, and possibly even desert elephants akin to the extinct species used by Hannibal's army in North Africa.[20] The mastery of the oxplow made farmers' labor efficient and allowed the full elaboration of crop strategies adapted to the highlands' island ecologies. This repertoire of short- and long-maturing crops allowed farmers to take advantage of localized conditions and also to make use of the short spring rains that occurred every few years in the highlands so that they could plant and harvest a second crop in a year.

Whether by local invention or early importation (more likely from South Arabia than the Nile Valley) the plow has been a part of the highland economy for two and a half millennia, including the epoch of the Kingdom of Daamat, an Ethio-Sabaen state that immediately preceded Aksum.[21] The oldest evidence for the oxplow in Ethiopia is a rock painting, dating from between 500 and 1000 B.C., which depicts two yoked oxen and a scratch plow remarkably similar to that used in Ethiopia in the late-twentieth century.[22] Early evidence of the plow in the highlands, however, is bound up with, and confused by, the interaction of Cushitic-speaking residents of the highlands and Semitic-speaking migrants from South Arabia. Though early schol-

Figure 3.6   Ethiopian plow, 1992. Photo by author.

arship tended to ascribe Aksum's innovations in technology, agriculture, and environmental management to South Arabian migrants, the best recent evidence supports the idea that plow agriculture preceded or at least coexisted with South Arabian influence. Archaeologist Joseph Michels has proposed a settlement model for Aksum based on surface evidence from Yeha (near Aksum) that he dates from 700 B.C. to 400 B.C. and associates with early South Arabian forms of irrigation and terracing. Only in the late–pre-Aksumite period (400–100 B.C.) did dryland oxplow cultivation not appear to be the predominate form of food production around Aksum itself. In the rock painting of the oxplow from Ambà Focadà the oxen are humpless, that is, are the older *Bos Taurus*, thus predating the arrival of zebu

Figure 3.7   Ethiopian plow, 1974. Photo by author.

cattle from the Nile Valley in small numbers around 1500 B.C. and then more widely around A.D. 670.[23]

There is little doubt that the labor efficiency of the plow and the sophistication of the cropping repertoire allowed the accumulation of surplus and fostered the creation of an elite political, priestly, and mercantile class that in the Aksum area evolved into a distinct urban population that occupied a number of sites around the main center at Aksum. This culture distinguished itself from neighbors in the Nile Valley and South Arabia by its language, as well as by its own style of pottery and cuisine. By the sixth and seventh centuries A.D. Aksum's urban center had reached its peak when the town covered an area of about 1 square kilometer.[24] The concentration of wealth in cities also enjoyed the benefits of a long-distance trade in luxury goods with the Red Sea, Nile Valley, and Eastern Mediterranean. Aksum's farmers were no doubt classic peasants whose food production in the form of taxes and tribute supplied the tables of urban dwellers and of the specialist trades (stone masons, iron workers, goldsmiths, religious specialists, scribes, etc). From its concentrated home population, Aksum was able to recruit armies and expand frontiers to the north, south, and west where it accumulated ivory, slaves, incense, alluvial

Figure 3.8 The Ethiopian plow, c. 500 B.C. (see upper right) in Ambà Focadà cave painting. From Antonio Mordini, "Un riparo sotto roccia con pitture rupestri nell'Ambà Focadà," *Rassegna di Studi Etiopici* 19 (1941).

gold, and other exotic trade goods in the form of tribute from peoples farther south.

Despite the evident sophistication of its elite material culture and extensive trade contacts, however, one could easily argue that the true genius of Aksum was its environmental management in smoothing and adjusting the vagaries of the seasons and its ecological setting. After all, the highlands' bimodal climate and sharply dissected plateau made transport and communication extraordinarily cumbersome. Periodic drought/famines offered further significant challenges to farmers and political leaders. Yet, Aksum's towns and capital city appear to have been substantially larger than the urban population centers of either the savanna or the Zimbabwe plateau and these urban sites needed a consistent supply of water, food, and labor to maintain the level of urban life evident in the archaeological record.

From the earliest evidence of pre-Aksumite times, it appears that the plow, iron tools, and concentrated labor (of humans and animals) had transformed the highlands' landscape. Stratigraphic sampling in

Eastern Tigray has shown a period of intense use of vegetative cover around the first century A.D. (about the time of Aksum's rise) followed by a period of regrowth around 1000 A.D. (about the time of Aksum's decline).[25] The plow and its need for open fields for annual crops and pastures for oxen over time created at Aksum a human landscape that over the course of a year cycled between images of a checkerboard of plowed fields, green sprouting crops, and golden ripening grain. Perhaps, like modern Ethiopia, trees and permanent vegetation were a feature of the towns and not the countryside, markers of concentrations of humans and not nature. As we learn from more recent evidence on the oxplow highlands, the lack of tree cover or deforestation per se did not necessarily lead to a decline in food production or in an environmental crisis (see Chapter 5).

Though the evidence is tentative, the ultimate secret to Aksum's elite urban culture may be elaborate skills of water management that balanced the needs of an urban population and agronomic challenges. Even in the best of years, the long dry season on the highlands inevitably creates problems of water supply and food during the months before the main fall harvest. In March and April springs dry up, oxen appear gaunt and bony, cows' milk output declines, and food stocks dwindle. How did ancient Aksum solve its daily problems of water supply when almost 75 percent of annual rainfall takes place within a three-month period (July–September)?[26]

A further water management problem existed even during the rains in that, while the town of Aksum itself sits on a rocky and well-drained terrain, its immediate vicinity, that is, the fields that supplied its granaries, straddles an expansive rich black vertisol ("cotton soil") plain. Vertisol is a black, fertile, and potentially rich soil for agriculture, but its behavior poses a peculiar set of problems to the farmer. When dry, vertisols form a deeply cracked clay surface too crusty and stiff to plow. When wet, vertisols first absorb water and can be plowed for a brief period; but then their surface quickly seals and becomes impervious to water, and the waterlogged soils drown germinating seeds. Once waterlogged, the vertisols become a mucky quagmire, a pudding impossible for humans and oxen to plow or even to move about on foot. Horses and mules sink up to their fetlocks. Vertisols thus pose the problem of how to drain them to allow fields to be plowed. Such an ecology challenged the farmer to devise a cropping system that tolerates waterlogging but still provides the year-round food needs of a dense human and livestock population and a non-farming urban elite. Yet the Aksumites could not afford to ignore those rich, black soils since they make up such a large portion of the tillable land: the Ethiopian highlands have 25 million hectares of these rich vertisols.

The Aksumite's choice to build their capital on top of a vertisol plain seems to indicate they considered it an asset rather than a liability. While we still await direct archaeological evidence of Aksum's ancient agricultural field systems, some tantalizing glimpses of Aksum's water management strategies exist. First, there is substantial evidence that Aksumites had developed an elaborate system of water harvesting that collected and stored rainwater in ponds and cisterns for use during the dry season. In Aksum town itself, the Mai Shum reservoir (described to tourists as Sheba's bath) has a large storage capacity of several thousand cubic meters, capable of providing an urban water supply for many months. A similar smaller cistern exists near a newly excavated Aksumite building, perhaps a palace, at Ona Nagast on the hill above the town. On the largest scale of all is the dam at Kohaito (northeast of Aksum on the main trade route) that is 76 meters long and 3 meters high.[27] Other water management sites, for irrigation and terracing, also exist in close proximity to Aksum town, especially at Abba Panteleon, which archaeologist Joseph Michels attributes to South Arabian influence. As in most of Ethiopia, terracing was a farming strategy employed primarily to conserve water and only secondarily to prevent erosion. These agricultural sites indicate the presence of a system of channeling water for irrigated agriculture (fruits and vegetables) and of preserving soil moisture for crops into the dry season. Such foods satisfied elite diets that must have valued fresh fruits, such as grapes, whose seeds have turned up in archaeological digs around the palace grounds. Aksum was a contemporary of Rome and may have shared the skills of the Mediterranean world in managing public water supplies.[28]

Aksumite achievements in water management must also have included field systems on the vertisol plain. To use their surrounding agricultural lands effectively, the Aksumites needed to apply their water-control skills to drain their soils, channeling away excess water as well as sowing crops tolerant of waterlogging. In the former case we have no direct evidence of what Aksumite farmers did in their vertisol fields, though we can observe what happens in highland farming systems with the same soil, oxplow equipment and seed materials. We can speculate that Aksum's farmers, like their descendants elsewhere in Ethiopia, may have built raised beds to channel water away and to create planting beds on ridges above the flooded channels. But we also can be certain that they employed and developed an ideal specialized cultigen: teff.

Teff is an endemic plant that evolved from human selection of native grasses of the *Eragrostis* genus. Its peculiar characteristics perhaps tell us more than archaeology alone about the ancient Aksumites. Teff is a grain whose kernel is only slightly larger than a grain of coarse sand. The delicacy of its size requires elaborate labor-intensive soil preparation before planting, often four to five passes of the plow to break down soil clods adequately. Its short wispy stalk is fragile and can easily lodge (collapse). Yet, its delicacy masks a toughness of spirit. Teff is highly resistant to drought, lasts longer in storage than any other cereal grain, and—most important for vertisol cultivation—it thrives on waterlogging that, as in paddy rice cultivation, suppresses weed growth.

Beyond its role in the Aksumite cuisine and its success on black soil, teff also added further benefits to make it an ideal match with the oxplow farming system: teff straw is the most palatable and nutritious dry season cattle fodder of any cereal crop. When stacked properly against post-harvest rains, it resists rot and provides a rich source of sustenance for oxen in the long dry season.[29] Teff straw also is likely to have contributed to housing Aksum farm and nonelite urban households since it is an ideal building material when mixed with clay (teff straw has been found in building bricks in Egypt). Teff's range of climate/altitude exactly reflects the conditions on the Shire plateau (1600–2200 meters).

Archaeological evidence confirms that Aksumites consumed teff; agronomic evidence also suggests that they were masters of its cultivation since there is a large arsenal of teff "land races" (subvarieties) in the Aksum area today. Aksum's varieties are highly attuned to local landscapes, doubtless the result of generations of highland farmers making careful seed selections for resistance to drought, flooding, disease, and pests. On those fields not planted in August in teff, farmers could wait until the rains had ended and the water standing in the fields had subsided and could plant lentils or peas that were able to germinate and mature on residual soil moisture after the rains had ended. Teff provided protein and carbohydrates as well as iron and other key vitamins; lentils, chickpeas, and peas offered protein to the diet and returned nitrogen to the soil.

By the management of soil, irrigation, and water harvesting, Aksum's food supply may have been able to overcome the harsh seasonality of pure dryland farming that has marked most of Africa's agricultural history. Elite urban diets could avoid the monotony of dry seasons without fresh fruit or green vegetables. The density of settlement on the highlands and the action of oxplows on the soil would have lent Aksum's highland landscape a far more human impact than was felt in either Great Zimbabwe or the Sahel.

Aksum's slow decline at the end of the first millennium A.D. has left few records. From its apogee in the sixth century to its decline in the tenth century A.D., the city of Aksum shrunk by half, from 100 hectares to 40 hectares. Aksum minted no coins after A.D. 680, indicating a change in its commercial economy. Ethiopian oral and written traditions cite a conquest of the city by Yudit, a "Falasha" queen who allegedly put the town to the torch late in the tenth century in what may have been some form of rural rebellion against Aksum's long standing hegemony. Whatever the origins of Yudit's challenge, she did not usurp its power or assume the throne. Yet the historical record is otherwise silent about the empire's fate. The expansion of Islam after the seventh century and the decline of regional trade in the Red Sea—trade depression is a pattern not uncommon in more recent times—is the most common explanation. After the eleventh century Aksum falls out of any historical record, though its symbolic role in Ethiopian Orthodox Christianity has continued. In the thirteenth century the center of political power in the highlands suddenly appears to the south under the control of an Agaw (Cushitic) dynasty that claimed legitimacy as an offshoot of the Solomonic lineage and Christian tradition of Aksum. Ethiopian capitals after the tenth century were either located at pilgrimage sites or elaborated military camps. Their locations have moved progressively further south from then into the twentieth century.

Some geologists and archaeologists cite environmental degradation as the root cause of Aksum's collapse as an urban-centered empire, emphasizing especially erosion as a potential source of economic and political decay because of the evidence of some erosion on hillsides near the town.[30] Yet, such evidence is not widespread, nor does the earlier evidence of fine-tuned management of natural resources around Aksum suggest Aksum residents' failure to understand the nuances of soil conservation necessary to sustain intensive agriculture.

Beyond a decline in regional trade and a military defeat, there is tempting evidence that Aksum's ecological conjuncture had come to an end. It is difficult to cite Aksum's urban decline as either a cause or effect of environmental changes, but the concentration of wealth from external trade during its development may have stimulated more careful management of intensive agriculture and water resources. Erosion evident around the town of Aksum may well have been a by-product of the dispersal of Aksum's urban population and not the cause.

With little direct evidence from historical meteorology for northeast Africa, it is nevertheless possible to surmise that the patterns of aridity that affected West Africa in the period after A.D. 1100 may have begun somewhat earlier in northeast Africa. Karl Butzer, a physical geographer who has worked extensively on Aksum argues that Aksum enjoyed a period of good spring rains during its ascendancy, an ecological benefit that later declined.[31] Indeed, a long epoch of reliable spring rains may have been critical to the water management of Aksum, leading not to the collapse of its agricultural system, since that system is likely the one still in place on the highlands, but to the failure of its urban center.

The ecological conjuncture that nurtured an urban political base in Aksum was a rare phenomenon indeed. From Aksum's departure until the twentieth century, political power on the Ethiopian highlands has been fundamentally a rural process.[32] Aksum was an interlude in a longer pattern of regional history made possible by human management and a conjuncture of nature.

ECOLOGY AND THE EVENTS OF HISTORY

Examples cited here highlight the difficulty and danger of associating a history of the environment and particular historical events. Environment is one of many factors that condition human behavior in the aggregate, but offer far too blunt an instrument to cite as direct causation. John Iliffe's thesis, therefore, is well taken in that it accepts the natural world as context and not as a discrete historical actor. Nevertheless, actions of the environment offer the contour and context for historical events and movements; they are an important component of the conjunctures that underlie economic change and the evolution of social institutions. Most important, the environment is not a fixed platform, but a shifting stage. Recent research on the role of environmental factors in historical change points to the fluidity of climate, disease vectors, and agriculture.

In the cases of African empire states reviewed here, caution in assessing historical causation is well advised. Historical environmental data up to the very recent past is only thinly documented, and local variations often may undermine sweeping generalizations. In all of the cases cited, the rise of centralized states in west, northeast, and south central Africa depended heavily on the contributions of international trade as well as local conditions that allowed the accumulation of wealth and emergence of central authority. Even in the first millennium A.D. the interaction of local, regional, and global scale are evident in the growth, elaboration, and decline of the Aksum empire, as with the Sahelian states in the following millennium. And the environmental stage is always shifting scenes in an infinite variety of combinations: geomorphology changes little during human epochs; climate and vegetation change seasonally and over long periods; disease is episodic.

But tying ecology to the history of political or economic events or social history is only one application for information on the environment of the past. Landscape history—what Africa looked like, when, and why—is also a part of human history with an aesthetic and not merely a pragmatic appeal. The role of climate, biodiversity, and the action of soils, as well as the overall appearance of landscapes, should be both a subject of Africa's history and a central component of how we reconstruct and define it. Chapter 4 examines more recent periods in African history where denser data are available. These data allow a more precise and nuanced reconstruction of African historical landscapes for the nineteenth and twentieth centuries when the social, environmental, and political effects of the expanding world economy reshaped and redefined Africa's ecology in both image and substance.

NOTES

1. George Brooks, *Landlords and Strangers: Ecology, Society, and Trade in West Africa, 1000–1630* (Boulder, 1993), 12–13. For further information on the genetics, distribution, and origins of African cattle breeds, see J. E. O Rege, G. S. Aboagye, and C. L. Tawah, "Shorthorn Cattle of West and Central Africa. I: Origin, Distribution, Classification, and Population Statistics," *World Animal Review* 78, no. 1 (1994): 2–13, and C. Meghen, D. E.

MacHugh, and D. G. Bradley, "Genetic Characterization and West African Cattle," *World Animal Review* 78, no. 1 (1994): 59–66.

2. Webb, *Desert Frontier*, 8.

3. For an example of this influence, see Ivor Wilks, "The Northern Factor in Ashanti History: Begho and the Mande," *Journal of African History* 2, no. 1 (1961).

4. Brooks, *Landlords and Strangers*, 174.

5. Webb, *Desert Frontier*, 4–5.

6. Ibid., 11.

7. Ibid., 10–11.

8. For West African jihads, see Marilyn Waldman, "The Fulani Jihad: A Reassessment," *Journal of African History* 6 (1966): 333 ff., and David Robinson, *The Holy War of Umar Tal: The Western Sudan in the Nineteenth Century* (Oxford, 1985).

9. Webb, *Desert Frontier*, 16, 26.

10. Great Zimbabwe refers to the primary stone-walled site itself, but here it is also used to describe the larger political and economic unit. For dates and a summary of research, see Graham Connah, *African Civilizations Precolonial Cities and States in Tropical Africa: An Archeological Perspective* (Cambridge, 1987), 194–95, citing T. H. Huffman, "Snakes and Birds: Expressive Space at Great Zimbabwe," Inaugural Lecture, University of Witwatersrand, 1981, 1.

11. For a study of the Portuguese and Sofala, see T. H. Elkiss, *The Quest for an African El Dorado* (Los Angeles, 1981); see also Connah, *African Civilizations*, 210, citing G. S. P. Freeman-Grenville, *The East African Coast: Select Documents from the First to the Earlier Nineteenth Century*, 2d ed. (London, 1975), 31, and T. H. Huffman, "The Rise and Fall of Great Zimbabwe," *Journal of African History* 12 (1972): 361–62.

12. For a thorough description of these ships and their history, see Erik Gilbert, "The Zanzibar Dhow Trade: An Informal Economy on the East African Coast, 1860–1964," Ph.D. dissertation, Boston University, 1997.

13. I. R. Phimister, "Pre-Colonial Gold Mining in Southern Zambezia: A Reassessment," *African Social Research* 21 (1976): 17.

14. Connah, *African Civilizations*, 188. D. N. Beach, *The Shona and Zimbabwe, 900–1850* (London, 1980), 28–29.

15. Carolyn Thorp, *Kings, Commoners and Cattle at Great Zimbabwe Tradition Sites* (Harare, 1995), chap. 3.

16. Connah, *African Civilizations*, 213.

17. Ibid., 200–201; R. Summers, *Inyanga: Prehistoric Settlements in Southern Rhodesia* (Cambridge, 1958), 137–41; and Phimister, "Pre-Colonial Gold Mining," 17.

18. For East African acquired immunity, see John Ford, *The Role of Trypanosomiasis in African History* (Oxford, 1971); Richard Waller, "Tsetse Fly

in Western Narok, Kenya," *Journal of African History* 31 (1990): 81–101; and James Giblin, "Trypanosomiasis Control in African History: An Evaded Issue?" *Journal of African History* 31, no. 1 (1990): 59–80.

19. Rodolfo Fattovich, "Archaeology and Historical Dynamics: The Case of Bieta Giyorgis (Aksum), Ethiopia," paper presented to European Association of Archaeologists, September 1997, p. 6.

20. Local tradition in Aksum claims elephants as a primary beast of burden. One stone engraving in Caleb's tomb depicts an elephant at work. Clearly, this issue awaits further archaeological research beyond the royal tombs and stele.

21. See Rodolfo Fattovich, "Remarks on the Pre-Aksumite Period in Northern Ethiopia," *Journal of Ethiopian Studies* 23 (1990): 1–33.

22. See James C. McCann, *People of the Plow: An Agricultural History of Ethiopia, 1800–1990* (Madison, 1995), 39–40; see also Antonio Mordini, "Un riparo sotto roccia con pitture rupestri nell'Ambà Focadà," *Rassegna di Studi Etiopici* 19 (1941): 59. It would be reasonable to speculate that the plow evolved from a type of digging stick attached to yoked oxen.

23. Rege, Aboagye, and Tawah, "Shorthorn Cattle," 4.

24. Fattovich, "Archaeology and Historical Dynamics," 9.

25. L. Brancaccio, G. Calderoni, M. Coltorti, F. Dramis, and Ogbaghebriel Berakhi, "Phases of Soil Erosion during the Holocene in the Highlands of Western Tigray: A Preliminary Report," paper presented at the 12th International Conference on Ethiopian Studies, East Lansing, Mich., 1994. These authors write of "deforestation," though the evidence suggests that montane or wooded grassland would be a more accurate description of the vegetative cover.

26. Tsegaye Wodajo, *Agrometeorological Activities in Ethiopia* (Columbia, Mo.: 1984), 10. Karl Butzer, "Rise and Fall of Axum, Ethiopia: A Geoarchaeological Interpretation," *American Antiquity* 46, no. 3 (1981): 471–95, believes that rainfall levels were much higher in Aksumite time than today, though the pattern of seasonality may be similar.

27. Butzer, "Rise and Fall of Axum," 479.

28. Michael DiBlasi of the University of Naples/Boston University team recently excavated a water channeling system, or drain, in the Aksumite palace at Ona Nagast, the first of its kind found in the area.

29. McCann, *People of the Plow*, 55.

30. Connah, *African Civilizations*, 90 also cites Butzer, "Rise and Fall of Axum," 471. See also Fattovich, "Archaeology and Historical Dynamism," 13.

31. Butzer, "Rise and Fall of Axum," 471–95.

32. See Frederick Gamst, "Peasantries and Elites without Urbanism: The Civilization of Ethiopia," *Comparative Studies in Society and History* 12 (1970): 373–92. Gamst justifiably discounts the Gondarine period when au-

thority, but not imperial power, rested in Gondar. See also Donald Crummey, "Some Precursors of Addis Ababa: Towns in Christian Ethiopia in the Eighteenth and Nineteenth Centuries," in *Proceedings of the International Symposium on the Centenary of Addis Ababa.* Addis Ababa, 1987. Crummey documents the existence of towns as markets and ecclesiastical foci but not as centers of political authority.

PART II

Africa's Historical Landscapes

Chapter 4

Desert Lands, Human Hands

In 1972–74 successive years of drought triggered a famine that swept across the African Sahel from Senegal to northern Ethiopia. Beyond its tragic human effects, the drought/famine also produced Africa's first media tragedy: images of cruel nature and human suffering appeared on television and in major print media. The most enduring image may be *Time* magazine's poignant full-page photograph of a desiccated and stiff carcass of a cow lying forlornly in a dry waterhole. Several years later a BBC reporter informed me that an ambitious news photographer had taken it upon himself to construct that riveting scene by dragging a dead bovine into the waterless depression because the surrounding landscape had already turned green from renewed rains.[1]

In 1987 the highly regarded PBS science program *Nova* broadcast a documentary film entitled "The Desert Doesn't Bloom Here Anymore." The film played on the public memory of the 1972–74 famine by presenting striking images of despoiled arid landscapes from West Africa and from the western United States. In both cases, the film argued, the cumulative weight of human action had transformed a delicately balanced natural order into environmental disaster. In the U.S. example mismanagement on corporate farms had led to declining productivity and poisoning of the soil; degradation in the African case, from Burkina Faso, resulted from human destruction of the vegetative cover, inducing both famine, and the ominous advance of the southern edge of the Sahara desert. Footage of the Burkinabe village showed bleak brown landscapes and conditions of stark poverty, images particularly effective with a viewing audience from green temperate latitudes. Computer-generated animation of a Sahelian landscape showed the disappearance of trees and the southern creep of the desert. The thesis was that the historical weight of human abuse under rapid population growth and the search for profit had created degraded land and made permanent changes in both landscapes and climate.

"The Desert Doesn't Bloom Here Anymore" had an ideological message—bad land management by humans creates deserts—hidden in its reportage. The argument was not just dogma but had a basis in science of the mid-1970s. In 1974 Joseph Otterman, an Israeli climate scientist who had worked in Israel's Negev desert, published his research hypothesis that posited a relationship between drought and the loss of vegetative cover. Specifically, he argued that bare, highly reflective soils in semiarid zones would increase surface "Albedo" (reflectivity) and reduce convective processes, thus resulting in decreased rainfall.[2] Otterman attributed vegetative change to overgrazing, though one could easily extend his thesis to include other economic activity that exposed the soil, such as tree cutting and annual cropping. Thus, in this view day-to-day human activity, especially tasks performed by women, were responsible for creating drought. This scientific hypothesis seemingly confirmed a popular suspicion in the West than African farmers and herders had brought on their own crisis.

A year after Otterman published his hypothesis, Jule Charney, an M.I.T. climatologist, proposed a similar cause for desertification that pointed a finger directly at local land use as the agent of climatic change. Charney argued that changes in vegetative cover could indeed increase aridity. He held that desertification was not the result of a single causal factor but an ever-tightening chain of events that began with the removal of plant cover, followed by a loss of the soils' moisture retention, resulting in a reduction of rainfall that, in turn, further reduced vegetation. With this science as a backdrop, the *Nova* film's pessimistic images of West African farmers closed the neo-Malthusian circle: under severe population growth African farmers had occupied fragile lands and their poor land management had pushed the Sahel into a slide toward desert.[3]

The science of the 1970s did not evolve from a tabula rasa but added to an existing set of conclusions about the presence and origins of desertification in Africa from an earlier historical epoch. Anthropologist Jeremy Swift has pieced together a fascinating chronicle of colonial land studies of the late 1930s that argued that the Sahara desert was advancing at an alarming and measurable rate. The most widely distributed and influential of these reports, by the colonial forester E. P. Stebbing, held that the desert's advance was the result of desiccation whose root cause was human misuse of resources. As a result of these findings from across colonial Africa, a French forester A. Aubréville coined the term *desertification* to describe the human-induced process of land degradation in the West African Sahel.[4]

Swift observed an important incentive for the spread of these stories within colonial governments and, later, international donors. Placing blame on African farmers and pastoralists, external agents justified colonial rule, central planning, and the urban control of rural resources. In the postindependence era, desertification, however vaguely defined and applied, as a justification of coercion and paternalism by urban elites, was already rather firmly in place by the time the African Sahel experienced the severe droughts of the mid-1970s.

Based on early colonial assertions, embryonic scientific hypotheses, and popular assumptions of the 1970s, the idea of human agency in African climate change diffused subtlely into international policy and media representations of the late 1970s, 1980s, and 1990s. In 1977, for example, the influential environmentalist group Worldwatch Institute and its director Lester Brown officially linked the Sahara's advance with African land use practices, such as overgrazing, increased cultivation, and firewood gathering. In that same year the United Nations Conference on Desertification asserted specific figures for desertification: 10 percent of the earth's surface was "man-made" desert and another 19 percent was under threat by human mismanagement. By the mid-1980s, linking African farmers and pastoralists with advancing desert was commonplace in the popular press, including such reputable organs as *National Geographic*, which projected the definitive aura of science and richly illustrated the degradation narrative.[5]

The Ethiopian famine of 1984–85 presented more immediate and more graphic images of Africans who were suffering amidst degraded landscapes. In 1991 the United Nations further added its bureaucratic imprimatur by officially defining desertification in anthropogenic terms as "land degradation in arid, semi-arid and dry sub-humid areas resulting mainly from adverse human impact."[6] International policy thereby placed Africa's environmental crises of the 1970s at the doorstep of Africans themselves, the result of both alarming rates of population growth and land degradation. This view was both historically based (even if wrongheaded) and judgmental: Africans were reaping the whirlwind of their past actions.

But what does actual evidence from the past contribute to this issue? In fact, empirical historical evidence and new findings in the science of climatology present a very different set of causes for the African Sahel's crises of the late-twentieth century. In this new research, there is evidence that human action and population growth indeed has changed the physical landscape by altering vegetative cover, but the meaning and direction of that change contradicts the human degradation narrative.

The evidence from new climatological and historical research questions the human hand in climatological degradation and adds new insights into Africa's history as a whole. As indicated in Part I of this book, Africa's Sahelian zone from Senegal in the west to Eritrea in the east has long been the stage for struggles between climate, water resources, and human strategies of political, social, and economic organization. Great African empire states from Ghana, Mali, and Songhay in the west, Bornu and Dar Fur farther east, and Nubia and Aksum in the Nile Valley have managed to link long-distance trade, urban settlements, and sophisticated polities with farm-level successes in food production within arid and semiarid zones. Human successes of the past, however, only underscore such achievements in overcoming the Sahel's environmental challenges.

In the late 1970s a new countervailing historical perspective on the Sahel's crisis of desertification came from climate historian Sharon Nicholson who combined the evidence from historical landscape descriptions, modern climate models, and research on lake levels from Lake Chad to reconstruct a history of desiccation of the Sahel.[7] From this evidence and rainfall data for the past 25 years, it is clear that the Sahel's climate patterns since 1970 represent the most dramatic sustained decline in rainfall ever recorded anywhere in the world. Nicholson's historical study plus George Brooks's and James Webb's work linking climate to historical processes put recent climate crises in perspective. Recurrent drought periods lasting one or two decades have been a persistent feature of the Sahel over the past five hundred years. The most recent period, however, has been remarkable in its low levels of moisture. The Lake Chad levels in particular indicate that the late twentieth-century desiccation is at least as severe as any in the last millennium.[8]

Pessimists, however, could also read Nicholson's 1978 evidence as confirming the cumulative nature of the climatic crises and the role of increasing population pressure and mismanagement as the cause. The current generation of climate science, however, casts an entirely different light on Nicholson's results, pointing to extra-African sources rather than human activity, as the cause of the Sahel's climatic catastrophy.

Climatologist Peter Lamb as early as 1978 had differed with the anthropogenic (human cause) thesis by pointing out the connection between oceanic conditions in the Atlantic and African climates.[9] By the early 1990s more science and consciously historical studies began to accumulate evidence to challenge the human degradation narrative for the Sahel and adjacent zones. In 1991 data from polar-orbiting satellites examined the movements of Sahelian vegetation over the 1980 to 1990 period in relation to annual changes in rainfall. This study showed that the Sahara's southern edge indeed had moved, but that it has fluctuated both north and south depending on the levels of each year's rainfall. It argued that previous evidence had been both overly localized and anecdotal.[10] Other work, this time at the British Meteorological Office, also reevaluated the evidence for vegetation's effect on the Sahelian climate. This research indicated that while land surface feedback can play a minor role in sustaining drought, ocean temperatures (especially the Hadley circulation in the North Atlantic) had a far greater role in determining rainfall patterns by affecting the position of the Inter-Tropical Convergence Zone (ITCZ). These larger oceanic conditions may result, in turn, from the effects of global warming that appear to differ between northern and southern hemispheres.

This new body of research also supports Lamb's evidence that points to a causal effect between lower ocean surface temperatures in the Atlantic north of the equator and higher ones to the south. There may therefore be some reason to believe that the Sahel's recent desiccation has been a product of an overall global warming that has occurred in the second half of the twentieth century. Desertification therefore seems attributable to global climatic processes more than local human action.[11] From this new perspective, farm-level actions along the edge of the desert may well be responses to long-term climate change rather than causes of desert encroachment. In this context a number of new studies of African land use have offered fresh insights into the relationship between African peoples, population, and landscapes. In other words, the African farmers in "The Desert Doesn't Bloom Here Anymore" were more likely to be victims of a changing environment rather than its perpetrators.

READING HISTORY BACKWARDS: COLONIAL ENVIRONMENTAL POLICY AND FOREST ISLANDS IN GUINEA, 1893–1990

> Never, I believe, has a year so dry occurred in Kissidougou. I am left to say that from year to year, rain becomes more and more rare. And this I do not find extraordinary—even the contrary would astonish me—given the considerable and even total deforestation in certain parts of this region. . . . The effects of this dewooding are disastrous; one would soon see nothing more than entirely naked blocks of granite. A region so fertile become a complete desert. Now there [remains] no more than a little belt of trees around each village and that is all.
>
> Director of Kissidougou Agricultural Research Station, 1893[12]

Debates over changing vegetation in West Africa have not been confined to the desert's edge. Early French colonial observers who passed through the forest/savanna mosaic—the transitional zone between humid forest and savanna in Guinea—remarked on the zone's economic potential and its "islands" of forest that, they concluded, were remnants of a full gallery forest left from an indeterminate past. French colonial policy from 1893 forward, and the Guinean national policies that followed independence in 1958, rested on the assumption that human population has broken down the area's "natural" vegetation and extended the savanna grasslands into historically forested zones. State policy for the twentieth century undertook to transform the behavior of local peoples to halt a process of deforestation that had never, in fact, taken place.

Ironically, just the opposite may have been true. In an insightful and innovative example of environmental fieldwork, James Fairhead, Melissa Leach, and a team of Guinean colleagues, have turned colonial history on its head, placing the historical evidence of the vegetative cover of Guinea's forest-savanna mosaic against a century of state-

Sheila Silver Library
Self Issue
Leeds Beckett University

Customer name: Mpunzi, Mazwi . (Mr)
Customer ID: 0331373965

Title: Green land, brown land, black land : an
environmental history of Africa, 1800-1990
ID: 1704845040
Due: 22/10/2015,23:59

Title: Methods in Human Geography: A Guide
for Students Doing a Research Project
ID: 170474184X
Due: 22/10/2015,23:59

Total items: 2
08/10/2015 18:40
Checked out: 3
Overdue: 0
Hold requests: 0
Ready for pickup: 0

Need help? Why not Chat with Us 24/7?
See the Need Help? page on the library
website: library.leedsbeckett.ac.uk

level policy based on a false but resilient degradation narrative that held that human habitation had destroyed the natural vegetation. Using archives, local interviews, aerial photography, and botanical fieldwork, Fairhead and Leach were able to reconstruct the history of the forest/savanna zone's vegetative cover and also trace the origins of a myopic set of assumptions that misinformed colonial policies on natural resources. Their research demonstrates in great detail the consistent misreading of the landscape history and how "reading history backwards" shaped policy from the French colonial state in the early twentieth century to Sekou Toure's First Republic's socialist programs after 1958.

Fairhead and Leach describe conflicting imagined landscapes: one in the colonial mind that saw human habitation as defiling a zone of continuous natural forest, and a contrasting view from local inhabitants that sees the forest mosaic as a *product* of concerted human action that has extended the forest through a specific set of land use practices.

Fairhead and Leach's study focuses on Kissidougou Prefecture set in West Africa's forest/savanna mosaic that lies between the Guinea savanna zone to the north and the tropical humid forest zone to the south. Rainfall in this region ranges between 1500 millimeters and 2100 millimeters per annum but is concentrated in a rainy season of 7–8 months. As in most African settings, the total amount of rainfall is less significant than the high degree of annual variation and the timing of the dry and wet seasons. In contrast to the northern edge of the Sahel's climate, this zone may indeed, since the mid-nineteenth century, be undergoing a rehumidification after a 200-year dry period.[13] Within these general conditions of geography and climate Kissidougou's farmers shaped a landscape whose origins and directions posed a puzzle that state officials have failed to solve for over a century.

Fairhead and Leach's major breakthrough in understanding the conflicting landscape histories in Guinea has been to demonstrate precisely how islands of forest evolved from human action within savanna ecology. For villagers of the savanna zone south of the Sahel, woodland belts are a logical and expected consequence of human habitation. That such a connection was logical and observable to Kissidougou villagers, but not to the putative ecologists of colonial rule, is the crux of the story.

Africa's savanna landscapes, like American prairies and Argentinean pampas, reflect the scourge of fire that sweeps across the flat grass-

Figure 4.1 Field-clearing fire in West African savanna/forest mosaic. (Courtesy of Combs, African Studies Program, University of Wisconsin-Madison.)

lands during the dry season. Fire burns off tall grass and creates space for new growth of palatable grasses, makes phosphorous available to new plants, and eliminates non-fire-resistant trees. Thatched roofs are vulnerable to fire, and thus new settlements tend to spring up and nestle near gallery forests and wetlands that offer some protection from dry season fires. Once settled, everyday village activities began with actions like gathering thatch for roofs and fences in the surrounding grasslands, reducing flammable materials, and creating a fire-free zone around each settlement. Village farmers then open fields for cultivation around village sites. The local farming techniques of mounding and incorporating organic matter into the soil retains moisture and encourages the growth of new tree species that otherwise have a difficult time surviving in hard-packed savanna soils. Rather than see their actions as harmful, farmers expect and generally find new woody vegetation to be a natural successor to abandoned fields.

Livestock also contribute to the building of forest islands on the savanna. Villagers tether cattle at the edge of forests away from entangling trees but close enough for them to feed on burnable grasses, thus "selecting" woody shrubs for further growth on newly manured soils. Farmers told Fairhead and Leach that large numbers of cattle on the savanna were not sources of degradation, but rather that their presence increased woody vegetation.[14] Over time, this induced succession of shrubs and fire-resistant trees gave way to species typical of gallery forests. In these shaded zones protected from fire, villagers planted shade-loving, commercially lucrative trees such as kola, banana, and, in later years, coffee. In turn, the forest islands became sources of valuable products such as seeds, nuts, basketry grasses, building poles, and medicinal plants—products particularly valuable to women. Villagers also revealed to Fairhead and Leach that these processes of forest building could be accelerated when needed, for example, to serve as defensive positions in the period before the Pax Colonia, or later as coffee plantations. New settlement sites thus sprung up adjacent to older forest settlements, extending the forest edge into the savanna. Mature gallery forests drew further human inhabitants because of their attraction as a site for wetland rice cultivation on swampy depressions that suppress weeds.

Fairhead and Leach make it clear that the savanna-forest ecology was far from the "natural" but degraded landscape that the first French visitors imagined. It was, in fact, a secondary vegetative cover induced by human action. Anthropologist A. Endre Nyerges has also described the social dynamics of forest vegetation (which he calls "the ecology of practice") on the Upper Guinea Coast. He examines the transformation of forest vegetation by the Susu people of Sierra Leone who practice swidden farming within the southern Guinea savanna/moist deciduous forest transition zone of West Africa, an area adjacent to Kissidougou (Fairhead and Leach's primary site).

Nyerges's fieldwork adds to the complexity of forest history in West Africa. He pays particular attention to the regenerative capacity of trees (coppicing) and the tensions between the changing social lives and resource management practice of Mande migrants into the forest zone in Sierra Leone. Nyerges's research notes that this zone was historically the site of a large population movement of savanna Mande peoples who migrated into the forest edge with the expansion of the Mali empire in the period 1100–1500 in search of charcoal for iron smelting and new lands for production of staple cereals. The dynamics of the settlement of the Mande-speaking Susu people in this ecology include local effects of the Atlantic slave trade (1500–1807), the eighteenth-century Fulbe rebellion against Mande rule, the settlement of an urban center at Freetown, colonial rule on the Upper Guinea Coast by French, British, and Portuguese forces, and the incorporation of the region into the world economy in the post–World War II period. His research addressed the questions of how these historical and demographic events acted on the ecology of the forest-savanna frontier. In this zone, he argued, there is considerable evidence of stability and an "applied restoration ecology" based on coppiced (regrowth of trees from cut stumps) woodlands managed by local populations.

Like Leach and Fairhead's case to the north, the Susu forest was secondary vegetation shaped by a particular set of human choices and patterns of use.[15] The patterns of Susu forest regrowth derived from an elaborate set of social institutions that reflected rank, gender, and the necessities of growing specific crops (especially New World crops such as chilies and groundnuts). Nyerges identifies two specific historical dynamics of forest change. The first was the historical movement of Mande iron smiths who moved into sparsely populated forest zones to make the charcoal necessary for their craft. The expansion of the iron trade in the second millennium reduced forest cover gradually until the nineteenth-century political dislocations and colonial rule disrupted local economies and, ironically, allowed forest cover to regenerate.[16]

The end of the ecology of charcoaling by Mande smiths led gradually to a new, more recent, dynamic. Nyerges's fieldwork in the early 1980s followed the use of forest plots by three different social groups: (1) elder men who control enough labor to cultivate old fallow upland rice, (2) younger men who control less of their own labor and who try to cultivate too-young fallow land near village sites, and (3) women who clear and then cultivate the elder men's former rice plots for chili peppers intercropped with groundnuts. Each of these groups seeks to maximize its control over labor, including their own, and apply it toward production of a specific crop—rice and chilies for men, groundnuts for women.

Nyerges argues that the effect of this status and gender division of labor on forest cover has shaped the landscape of the forest/savanna frontier over time. Left alone, tree stumps in old fallow rice plots send out new shoots and regrow (coppice) into secondary forest over time, as happened historically with the decline of local iron production in the early to mid–nineteenth century. Recent local farming practice, however, deflects this succession by laying open forest sites to invasion by fire-tolerant savanna trees and grasses. Specifically, the actions of younger men who are seeking to build their own farms and household labor supplies have tended to establish temporary farms on too-young plots that require the burning of grass cover and the destruction or delay of the coppicing of trees. The result in Nyerges's view can be a temporary or permanent change of forest to savanna or tree savanna.

The key ingredient of historical forest change would seem to be the introduction of such New World crops as groundnuts (*Arachis hypogea*) and chilies (*Capsicum spp.*) cultivated for the market. The adoption of these crops in the 1830s broke the thirty-year fallow cycle of rice production that had allowed the full regrowth of moist forest. Production of chilies and groundnuts on former upland rice plots by using women's labor and the interruption of long fallow by younger labor-starved households threatened the stability of forest succession. Thus, both human social institutions and new crops imposed new demands on the forests' "natural" succession.

Is this research on change in Upper Guinea forests in conflict with the work of Leach and Fairhead who argue for human action as the cause of forest regrowth? In both cases human activity (of Mande peoples) has transformed the forest cover. Fairhead and Leach see an increase in tree cover onto former savanna while Nyerges sees the loss of moist forest to tree savanna or savanna. Both approaches would agree that Upper Guinea forests are *socially* derived but seem to differ on the net result of such human action on the extent of forest cover. Nyerges in his more recent work (1996), in fact, acknowledges that forested sites have emerged from former savanna lands where human settlement produced "an accumulation of human bodily waste, fuelwood ash, food wastes, and the residues from the cultivation of kitchen gardens." Ultimately, he defends his own evidence of ecology change and forest loss by arguing that the imposition of a cash economy in Susu forests created a specific conjuncture that interrupted the process of "restoration ecology" just as other local conditions may have promoted forest expansion.[17]

West Africa's forest cover in the twentieth century was thus the outcome of particular local histories. In sum, evidence from the Upper Guinea forest argues strongly against the Amazonian model that implies a progressive, unidimensional loss of forest.[18] Ironically, however, the evidence for a human role in forming the changing vegetative landscapes at the southern end of the savanna contrasts with the preponderant role of physical forces—oceanic effects—in the movement of vegetation at the Sahel-Saharan edge. At least one more case, this time in East Africa, sheds light on the role of human activity, especially population growth, in Africa on the degradation of grassland landscapes.

CAUSE AND EFFECT: POPULATION AND PLANTS IN MACHAKOS, KENYA, 1939–90

In 1937 C. Maher, a colonial ecologist, described the landscape of Machakos in eastern Kenya. His description reflected a well developed colonial narrative of African degradation:

> The Machakos Reserve is an appalling example of a large area of land which has been subjected to uncoordinated and practically uncontrolled development by natives whose multiplication and the increase of whose stock has been permitted, free from the checks of war and largely from those of disease, under benevolent British rule.
>
> Every phase of misuse of land is vividly and poignantly displayed in this Reserve, the inhabitants of which are rapidly drifting to a state of hopeless and miserable poverty and their land to a parching desert of rocks, stones, and sand.[19]

Maher's grim observations were not isolated. Through the 1920s and 1930s, British colonial officials who observed the Machakos district in eastern Kenya described a landscape devoid of natural resources, an area virtually denuded of vegetation and in danger of completely losing its topsoil to massive gully erosion. Like their French counterparts in Guinea of the same period, British commissioners saw the problem as essentially human induced. The 1929 Hall Commission reported morosely that "it is not too much to say that a desert has already been created."[20] In 1937 two field reports confirmed that assessment and firmly placed the blame for the denuded hillsides and valleys on the local Wakamba people, their crops, and their herds. C. Maher, author of one of the reports and the quotation cited above, listed the chief causes, in descending order of importance, as deforestation, overstocking, cultivation of slopes, overcultivation, plowing, increases in cultivated area, road drainage, and livestock damage.[21]

Colonial records, in fact, consistently describe a terrain heavily smitten by a growing local population whose hearths, herds, and fields had rapidly transformed the look of the land over the first third of the twentieth century. But what did these changes mean about the area's future? The British colonial view was clear: Machakos would soon be a permanently degraded desert, caused as much as anything by the growing numbers of Wakamba who sought to eke out a living there by grazing their herds and avoiding raids from nearby Maasai. Yet, by the early 1990s, Machakos was far from a desert. On the contrary, though eastern Kenya had one of the world's highest population growth rates, its farms were productive agriculturally, and Wakamba homesteads enjoyed a green, almost lush, cover of trees, shrubs, and garden crops. What happened? Did British observers lie about what they saw, or did their policy recommendations save the day and stop the degradation in its tracks? Or was their environmental crystal ball just wrongheaded? It may be closest to the truth to say that colonial officials accurately depicted what lay before them in 1937 but only described a snapshot of a moving train, an ecological transformation in process, failing to discern both its movement and its direction. Misreading the nature of the changes underway, they woefully misjudged the landscape's dynamics and potential in a way that echoes the views of French colonial observers of West Africa's forest/savanna mosaic.

It is possible to make this judgment in hindsight because of the research of a hardworking team of environmental historians and ecologists led by Michael Mortimer and Mary Tiffen from Britain's Overseas Development Institute (ODI). The team, including environmental scientists from the University of Nairobi, sought to challenge neo-Malthusian wisdom about the effects of population growth on food supply, natural resources, and human welfare. Their goal was to reconstruct the environmental conditions of the 1930s and lay them alongside the evidence of the state of Machakos's natural resources of the 1990s. Along the way they hoped to describe the process by which Machakos's high population growth coped with its stark landscape and transformed it.

The Mortimer-Tiffen team employed a number of innovative research techniques: comparative photography, archival research, rangeland succession assessments, aerial photography, and field interviews.[22] The result was the most comprehensive reconstruction to date of the shaping of an African landscape under human management. Their research not only directly addresses neo-Malthusian assumptions about the negative effects of population growth but also offers further insights into the issue of desertification. In fact, their research allows us to witness a process of landscape change from grassed shrubland to dense wooded shrubland capable of supporting livestock, cropping, and the resource needs of a dense human population.

Machakos district lies to the east of Nairobi in eastern Kenya. It consists of a series of hills connected by plains and occupies an area about twice the size of Rhode Island. Rainfall varies between 1100 and 500 millimeters per annum but is both bimodal and highly variable from year to year. It is typical of Africa's fragile, semiarid zones that have in recent years received large movements of populations who are seeking new agricultural lands and pasture. Moreover, the Inter Tropical Convergence Zone (ITCZ) was a fickle visitor, its northern movement tending in some years to moisten only south-facing slopes, leaving north-facing slopes in a rainfall shadow, that is, drought prone.[23] For this reason only about a third of the area receives the 750 millimeters of rainfall necessary for crop production, and only a small proportion of Machakos can expect more than 250 millimeters of rainfall (the bare minimum for maize production) in six or more years out of ten.[24] Over the last hundred years, in fact, Machakos has experienced a number of droughts, the most severe of which occurred in the late 1890s and in the mid-1970s, roughly similar to patterns in the Sahelian zones of the Horn of Africa to the north.[25]

At the end of the nineteenth century, Wakamba people in Machakos had settled themselves in the hills, using the grass-covered plains as grazing land that they alternately shared and struggled over with the Maasai. While Wakamba had cleared most hill slopes for cultivation, there was little evidence of erosion. Early impressions of Machakos were conditioned, however, by effects of the 1898–99 drought and the 1909 rinderpest epizootic that reduced livestock numbers and concentrated the human population on the better-watered southern hill slopes. Enthusiastic reports found in traveler's descriptions that commented on the "wonderful grass to be seen on the caravan route" reflected postcrisis conditions with diminished livestock herds rather than a static state of nature.[26]

According to Wakamba accounts, the arrival of a British administration created new conditions of human settlement that in turn changed the landscape, beginning the long transformation of soils and vegetation. British colonial rule provided security from Maasai raids, allowing Wakamba cultivators (mainly women) to extend their fields safely onto the lower slopes and plains. Herders burned the tall plains grass to clear land for fields and to obtain fresh growth for livestock. Shrub trees provided fuel and building materials for the human and livestock population expansion that inevitably followed the period of extended drought.

The colonial government gazetted the area in 1906, fixing the boundaries of Wakamba cultivation and pasture. Machakos thus found itself bounded on the north-west by White farms, on the north-east by Crown land, and by tsetse-infested bush on the southeast. By the mid-1920s, Wakamba herds had therefore lost grazing areas that had sustained livestock during droughts and suffered reduction of land available for shifting agriculture. Serious droughts occurred in 1907-8, 1910, 1913-14, and 1917-18. By 1922 the government's annual report stated that Wakamba cattle were dying "literally from starvation."[27] Restrictions on pasture and failing rains began in the 1920s to create the barren, eroding landscape that British officialdom observed in detail and blamed squarely on the Wakamba. By 1927 the Machakos district commissioner reported: "Since 1917 the reserve has become desiccated beyond all knowledge. Large areas which were good pasture land, and in some cases thick bush, are now only tracts of bare soil."[28] Moreover, from 1931 to 1936 six moderate or severe droughts compounded the effects of intensified pressure on natural resources. Given the recent remote-sensing evidence of the effects of drought on vegetation cited earlier in this chapter, it is not surprising that colonial officials found the land denuded and soil increasingly disturbed and eroded. The 1937 reports (quoted above) that documented soil erosion thus were commenting on a threefold set of factors affecting human land use, vegetation, and soils: (1) the hemming in of Wakamba herds, (2) the first phases of land occupation on the lower slopes and plains, and (3) the effects of drought on vegetative cover. Maher and his fellow colonial officials, however, concluded myopically that the primary cause was Wakamba misuse of the land and the likely result was irreversible degradation and desertification. Their conclusions on the human causes of land degradation, in fact, stemmed less from observations of Wakamba farming and herding systems than from the narratives coming from the American Dust Bowl experience, though in the African case overstocking of cattle rather than arable agriculture was the primary object of concern.[29] Government policies in the 1950s and postindependence continued

Table 4.1
**Changes in the Vegetation Structure of Kathonzweni and Ngwata
between 1960 and 1990**

Area and Year	Physiognomic classification	Life-Form (% cover)		
		Trees	Shrubs	Grasses
Kathonzweni				
1960	grassed shrub	2	39	59
1978	grassed shrub	9	35	51
1990	dense wooded shrub	15	38	29
Ngwata				
1960	grassed shrub	1	37	62
1978	grassed shrub	6	45	19
1990	dense wooded shrubland	11	62	18

the degradation narrative of the 1920s and 1930s when confronted with alarming evidence of rapid population growth. From this perspective a Malthusian apocalypse seemed inevitable.

This view that has dominated the literature of colonial ecology as well as postindependence development policy has recently come under empirical scrutiny by historically sensitive ecologists. Dr. Kassim O. Farah of the University of Nairobi has examined the evidence for changes in the natural vegetation of Machakos for the 1930–90 period and has challenged the assertion that degradation had taken place.[30] As a point of departure Farah argues that by 1930 the Kamba Reserve (Machakos) already consisted of managed rather than natural vegetation. Then, using aerial photographs from 1960 and 1978 and a field inventory of vegetation undertaken in 1990, Farah set down a longitudinal base from which he could examine vegetative change in Machakos.

Farah concluded that there was a noticeable increase in woody vegetation in the sample sites where in each case the vegetation had changed from grassed shrubland to dense wooded shrubland (Table 4.1). This shift in vegetative structure between 1960 and 1990 likely replicates a similar process that took place in the 1930 to 1960 period when Wakamba cattle grazed in the confined zone of the Reserve.

The continued grazing of cattle on grasses tends over time to encourage the growth of woody plant species. Yet, Farah concluded that the most significant factor in the shift from grassland to dense shrub woodland was the role of fire. Wakamba used fire historically in range management to promote the growth of new grasses and to control woody shrubs. Colonial and Kenyan governments prohibited the use of fire and thus inadvertently played a major role in increasing the woodiness of grazing lands. So, there has been an overall decline in grazing capacity (grass replaced by woody species) and a net increase in woody shrubs, including trees. This change is in theoretical terms at least a higher successional status. Though some loss of grazing capacity occurred, Farah argues, the levels of decline would be important only for the needs of commercial beef production and not for smallholder integrated farms.

The results of the Machakos study therefore turn the Malthusian thesis about the effects of increased habitation 180 degrees: increased human (and livestock) settlement actually increased plant cover and arrested earlier forms of erosion that, it appears, may have been only a transitional stage in the succession of landscape change rather than a terminal condition leading to desertification.

That African landscapes and vegetation change over time should not be in doubt. What new research allows us to see are the complex dynamics of that change. A further example from East Africa, in this case the Serengeti plain, illustrates the chronology of change on Africa's most publicized piece of African nature. Holly Dublin, an ecologist who has worked in East Africa since 1981, examined the botanical and historical records of the Serengeti/Mara ecosystem over the course of the twentieth century. Her research challenges the conventional wisdom that East Africa's savanna woodland ecosystem was a stable "climax" community. In fact, Dublin shows that over the course of the twentieth century the Serengeti/Mara landscape has had three distinct vegetative shifts within the period of a hundred years for which there are clear visual and documentary records. These records include travel narratives from European observers, landscape photographs, and aerial photographs that allowed a detailed quantitative comparison of woody cover for the 1950–82 period.[31]

Dublin's research shows, in sum, that the Serengeti/Mara landscape shifted from an open grassland in the 1880s to a dense woodland in the 1930s and then back again by the 1960s when the Serengeti achieved the nadir of its international celebrity as a pristine and protected African wildlife landscape. To account for these changes Dublin examines the tightly intertwined effects of fire, elephants, and human beings. The evidence from the Serengeti on human agency, however,

does not suggest a direct link between human population and deforestation and/or desertification, but rather describes a complex set of
relationships within the Serengeti biome resulting in landscape
change.

Maasailand, which included the Serengeti, underwent a set of
broadly based environmental shocks that affected much of East Africa
in the final decade of the nineteenth century. In September 1890 the
livestock disease rinderpest arrived on the Serengeti plain, killing as
many as 90 percent of the Maasai cattle population within a few
months. Not only cattle but also gregarious, herd-forming wildlife,
such as wildebeest and buffalo, died in large numbers. Maasai pastoralists who depended on cattle either died or abandoned the region.
Many local people became *dorobo* (hunters and gatherers) as a strategy
to survive. The transformation of the region's ecology and economy
brought other changes in wildlife distribution, that is, the reduction
of the elephant population through the expansion of ivory exports
into the revived Indian Ocean trade from Zanzibar and the presence
of desperate pastoralists seeking to survive.

By 1900 the combined effects of disease, famine, and expanding
ivory markets had left the Serengeti plain devoid of most of its previous fauna, human and otherwise. Travelers to the area in the first
decade of the twentieth century described "undulating grasslands,
lovely fertile country which seemed almost uninhabited," or "a broad
plain of park-like country, fine grazing land, studded with the occasional yellow-barked *Acacia*."[32] Others remarked on the peculiar absence of species like elephants and buffalo.

Three decades later these changes in faunal demography yielded a
profound change in the Serengeti landscape. Hilltop *croton* thickets
(*Acacia syal* and other acacia species) replaced the open grasslands,
and hunters reported abundant leopard populations (which prefer
wooded to grassy settings).[33] Because of the ideal wooded habitat, the
Serengeti also became a prime area for trypanosomiasis (sleeping sickness) and its vector the tsetse fly. Wildlife that had survived rinderpest
developed immunity to it and became primary hosts for trypanosomes. The colonial government's attempts to control tsetse flies failed
and anti-tsetse programs collapsed in the early 1960s after massive
population relocations that cleared the Serengeti of its human inhabitants.[34]

The shift of Serengeti woodlands to the open grassland setting of
the 1980s and 1990s took place as a historical conjuncture of natural

Figure 4.2   Acacia woodland, east of Serengeti, 1974. Photo by author.

and indirect human forces. Fire has long been an important tool in African range management, hunting, and agriculture. It was a tool of local peoples to manage vegetation or promote soil fertility or improve pasture. Holly Dublin has argued forcefully that unusually high rainfall in the early 1960s resulted in a dramatic increase in grass growth that provided fuel for hot and particularly destructive bush fires. Fires were also set by park authorities as part of fire management schemes, or by accident through local honey hunters, European hunters, or lightning strikes. With increased grass growth and wooded areas for fuel, fires in the late 1950s and early 1960s destroyed much of the vegetative growth associated with the Serengeti/Mara area since the 1930s.[35]

To this factor Dublin has added the effects of rising human populations at the edge of the Serengeti that drove elephants and grazing wildlife into the protected reserve. The increased density of the elephant population also contributed to the loss of the wooded landscapes. Elephants feed on acacia leaves and bark and a single elephant can obliterate a 5-meter-high tree in less than fifteen minutes. Fires and the reduction of dense wooded areas also attracted grazers to post-fire pasture and helped push back the woodlands' hegemony. According to Dublin's photographic evidence, between 1950 and 1982

Figure 4.3 Open grasslands, east of Serengeti, 1974. Photo by author.

over 95 percent of the Acacia woodland in Serengeti had become grassland. Of this change, 65 percent took place between the years 1961–67 (i.e., the years following the highest rainfall amounts). The serendipity of nature and the human intervention to "protect" the Serengeti thus returned a grassland to a woodland, and back again.[36]

CONCLUSION

There is a general consensus that the degradation of an ecosystem describes a stage at which it loses its resilience, resulting in a permanent decline in its productive capacity to support humans nutritionally and economically. Over the course of the twentieth century, external observers have characterized African landscapes—soils, rangelands, crops, and even its climate—as degrading or degraded. One of the most graphic of these narratives has been an allegation that African lands are becoming desert because of human actions in promoting erosion, deforestation, and loss of vegetation through overgrazing and bad agricultural methods. The evidence presented in this chapter seriously challenges this thesis in a number of different geographical and ecological settings.

That African landscapes have changed over the course of the twentieth century is not in question. It does seem clear in this chapter and the next, however, that the directions of change are neither fixed nor unidirectional. Moreover, the historical conjuncture of wider global forces in climate or faunal change, or the serendipity of a lightning strike, may override or redirect the actions of African farmers and pastoralists or conservationists in determining the productive capacity of land, the physical condition of land, and its human and plant overlay.

Moreover, African landscapes at present or in historical settings are in a state of movement. "Reading" a landscape therefore means getting right the dynamics and the direction of its movement as well as describing what meets the eye in a given moment in time. The cases of Kissidougou and the Serengeti point to the dynamism of landscapes themselves and the forces that shape them. Across Africa, the period of the 1930s appears to have been crucial in setting a story of degradation and prescribing the world public's image of what African landscapes ought to look like. Subsequent efforts at conservation often were attempts by human agents to freeze the landscape's dynamism and achieve a scene that conforms to prevailing ideas about Africa's "natural" state. Evidence from the Sahel, Guinea's forest-savanna mosaic, and the Serengeti/Mara grasslands suggest that humans in nature indeed change their surroundings but in ways that are worthwhile to observe but difficult to predict. The next chapter offers further evidence (from northeast Africa) that popular images of African landscapes are often misguided, or at least misleading.

NOTES

1. Patrick Gilkes, personal communication. Gilkes was then editor of the BBC's *World Service's Focus on Africa* program.

2. J. Otterman, "Baring High-Albedo Soils by Overgrazing: A Hypothesized Desertification Mechanism," *Science* 186 (1974): 531–33.

3. J. G. Charney, "Dynamics of Deserts and Drought in the Sahel," *Quarterly Journal of the Royal Meteorological Society* 101 (1975): 193–202. See also Mike Hulme and Mick Kelly, "Exploring the Links between Desertification and Climate Change," *Environment* 35 (July–August 1993): 5–11, 39–45.

4. Jeremy Swift, "Desertification: Narratives, Winners and Losers," in *Lie of the Land*, ed. Leach and Mearns, 73–90.

5. E. Eckholm and L. R. Brown, *Worldwatch Paper* No. 13, Washington, D.C.: Worldwatch Institute, 1977, p. 1; *United Nations Conference on Desertification, Round-Up Plan of Action and Resolutions* (New York, 1977), 2, cited in Swift, "Desertification Narratives, Winners and Losers," 80; W. S. Ellis, "Africa's Sahel: The Stricken Land," *National Geographic* 172 (August 1987): 141–79.

6. United Nations Environment Program, *Status of Desertification and Implementation of the UN Plan of Action to Combat Desertification*, UNEP/GCSS.III/3. Nairobi: United Nations Environment Program, 1991.

7. S. E. Nicholson, "Climatic Variations in the Sahel and Other Africa Regions during the Past Five Centuries," *Journal of Arid Environments* 1 (1978): 3–24.

8. Hulme and Kelly, "Exploring the Links," 10.

9. Lamb, "Circulation Patterns," 240–51.

10. Compton Tucker, Harold E. Dregne, and Wilbur W. Newcomb, "Expansion and Contraction of the Sahara Desert from 1980 to 1990," *Science* 253 (July 1991): 299–301. See also "Satellites Expose Myth of Marching Sahara," *Science News* 140 (July 1991): 38.

11. Hulme and Kelly, "Exploring the Links," 40–41. For definition of degradation, see Jerold L. Dodd, "Desertification and Degradation in Sub-Saharan Africa," *Bioscience* 44 (1994): 28–33.

12. Nicolas, "Etat de cultures indigènes," August 1914, Archives Nationales de la Republique de Guinee, Conakry. Cited in Fairhead and Leach, *Misreading the African Landscape*, 240.

13. Fairhead and Leach, *Misreading the African Landscape*, 50. They cite S. E. Nicholson, "The Methodology of Historical Climate Reconstruction and Its Application to Africa," *Journal of African History* 20, no. 1 (1979): 31–49, and George E. Brooks, "A Provisional Historical Schema for Western Africa Based on Seven Climatic Periods," *Cahiers d'Etudes Africaines* 101–102 (1986): 1–2, 43–62.

14. Fairhead and Leach, *Misreading the African Landscape*, 225–30.

15. A. Endre Nyerges, "Deforestation History and the Ecology of Swidden Fallows in Sierra Leone," Boston University African Studies Center

Working Paper No. 185, 1994, pp. 3–4, and Oliver Rackham, *Ancient Woodland: Its History, Vegetation and Uses in England* (London, 1980).

16. A. Endre Nyerges, "Ethnography in the Reconstruction of African Land Use Histories: A Sierra Leone Example," *Africa* 66, no. 1 (1996): 134. Nyerges's chronology relies heavily on Brooks, *Landlords and Strangers*, passim.

17. Nyerges, "African Land Use Histories," 138–40.

18. For a description of and challenge to the Amazonian model, see, E. F. Moran, "Deforestation and Land Use in the Brazilian Amazon," *Human Ecology* 21, no. 1 (1993): 1–21.

19. D. B. Thomas, "Soil Erosion," in *Environmental Change and Dryland Management in Machakos District, Kenya, 1939–90: Environmental Profile*, ed. Michael Mortimer, ODI Working Paper No. 53, December 1991, pp. 27–28.

20. Thomas, "Soil Erosion," 27.

21. Maher quoted in ibid., 28.

22. Michael Mortimer, Mary Tiffen, and Francis Gichuki, *More People, Less Erosion: Environmental Recovery in Kenya* (London, 1993), 25–30.

23. For a worldwide description of these lands, see Michael Glantz, ed., *Drought Follows the Plow* (Cambridge, 1995).

24. S. K. Mutiso, Michael Mortimer, and Mary Tiffen, "Rainfall," in *Environmental Change and Dryland Management*, ed. Mortimer, 3–4, 13.

25. Mutiso, Mortimer, and Tiffen, "Rainfall," 19. The authors looked for patterns within Machakos and not wider connections to African climate overall.

26. J. R. Perbody, *Machakos District Gazetteer*, Machakos District Office, Ministry of Agriculture, 1958, quoted in Thomas, "Soil Erosion," 26.

27. Thomas, "Soil Erosion," 26, quoting the Annual Report, Ukamba District, 1922.

28. Perbody, *Machakos District Gazetteer*, 1958 cited in Thomas, "Soil Erosion," 27.

29. David Anderson, "Depression, Dust Bowl, Demography, and Drought: The Colonial State and Soil Conservation in East Africa during the 1930s," *African Affairs* 83 (1984), 321–43.

30. Kassim O. Farah, "Natural Vegetation," in *Environmental Change and Dryland Management*, ed. Mortimer, 51–66.

31. Holly T. Dublin, "Dynamics of the Serengeti-Mara Woodlands: An Historical Perspective," *Forest and Conservation History* 35, no. 4 (October 1991), 169–78.

32. Mary A. B. Buxton, *Kenya Days* (London, 1927), 67; R. B. Woosnam, "Report on a Search for Glossina on the Amala (Engabei) River, Southern Maasai Reserve, East Africa Protectorate," *Bulletin of Entomological Research*, 4 (1913): 275, quoted in Dublin, "Serengeti-Mara Woodlands," 173.

For the effect of the 1890–92 crisis on the Maasai, see Richard Waller, "Emutai: Crisis and Response in Maasailand, 1883–1902," in *The Ecology of Survival in Northeast Africa*, ed. David Anderson and Douglas Johnson (Boulder, 1988).

33. Dublin, "Serengeti-Mara Woodlands," 173.

34. For trypanosomiasis policy, see Kirk Hoppe "Lords of the Flies: Environmental Images and Social Engineering in British East African Sleeping Sickness Control, 1903–1963," Ph.D. dissertation, Boston University, 1997.

35. Dublin, "Serengeti-Mara Woodlands," 174.

36. Ibid., 175.

Chapter 5

A Tale of Two Forests:
Narratives of Deforestation in Ethiopia,
1840–1996

One tragic example of the loss of forests and then water is found in Ethiopia. The amount of its forested land has decreased from 40 to 1 percent in the last four decades. Currently, the amount of rainfall has declined to the point where the country is rapidly becoming a wasteland. The effects of the prolonged drought that have resulted have combined with the incompetence of its government to produce an epic tragedy: famine, civil war, and economic turmoil have wreaked havoc on an ancient and once-proud nation.

Senator Albert Gore
Earth in the Balance (1992)[1]

Albert Gore's conflation of deforestation with drought, famine, and historical tragedy was a significant historical statement, though not for its truth or innacuracy. On the contrary, his assertion about forests in Ethiopia marks the steady passage of a single environmental story or narrative—the past state of Ethiopia's forest—from the status of unfounded conventional wisdom to a statement of fact by one of the world's most prominent environmental policy makers. Indeed, there is a strange harmony between the scholarly and public policy opinions on the historical condition of Ethiopia's forests. For many, Ethiopia's trees have become symbolic of a general theme of natural resource degradation in the modern developing world, a narrative that uses embedded assumptions about history as a pretext for environmental policy.

What is the implicit argument about the history of nature and human action in this degradation narrative? Is there an empirical foundation for the precise, didactic deforestation figures offered by Gore? From where did those deforestation statistics derive? This chapter will examine the empirical historical evidence of forest cover in modern Ethiopian history and indirectly will explore both the nature of conventional wisdom about the African environment and the meaning of historically based degradation narratives.

The public and scholarly assertions about the historical and present state of Ethiopia's forest resources are a specific and testable case of a degradation—or declension—narrative. The story of Ethiopia's diminishing forests is not just a pearl of didactic wisdom for policy makers but also a politically visible judgment about Ethiopia's past and how its people have allegedly mismanaged resources. A history of these putative facts about Ethiopia's forest cover and an exploration of their accuracy will illustrate the power of apocryphal environmental narrative and will demonstrate the value of empirical research in environmental history. Such a search of the historical record also reveals the existence of earlier narratives that ascribed quite different meanings to Ethiopia's forest cover.

ORIGINS OF A NARRATIVE: ETHIOPIA'S 40 PERCENT FOREST COVER

Albert Gore's comments on Ethiopia's vegetative cover build upon a thirty-year history of involuted citations of unsubstantiated "fact" in the absence of field evidence. A short history of Ethiopia's alleged historical 40 percent forest cover is in order here. The deforestation figures published in *Earth in the Balance*, with some variation in baseline year, have their origins in a less-public, but nonetheless ubiquitous, set of putative facts circulating within the development and scholarly community that has seen deforestation as an indicator of Ethiopia's environmental history and a harbinger of a bleak future.

The first published reference to a figure of Ethiopia's 40 percent forest cover appears in *Agriculture in Ethiopia*, a 1961 FAO publication by H. P. Huffnagel.[2] Huffnagel's otherwise valuable survey of

Ethiopia's agricultural resources states that by 1960 the 40 percent forest cover present in 1900 had diminished to just 4 percent but cites no source for the assertion. His source may well have been personal communication from Professor F. von Breitenbach, the founder of Ethiopian forestry studies, whose 1962 article "National Forestry Development Planning" asserted that "the Ethiopian high plateaux were, in ancient times, almost completely covered by more or less dense forest." Von Breitenbach hypothetically estimated that Ethiopia's ancient forest cover at 37 percent, a figure he based, it seems, on those land areas with a *climatic potential* for forest cover.[3] In other words, von Breitenbach rooted his statement on forest cover on rainfall only, that is, he had no field surveys or physical evidence whatsoever. Sixteen years earlier W. E. M. Logan's book *An Introduction to the Forests of Central and Southern Ethiopia* (1946) had speculated on levels of deforestation based on the presence of remnant high-forest species, estimating that 5 percent of Ethiopia was forested in 1946.[4] Thus Huffnagel's precise figures seem to be a conflation of estimates and speculation of forest specialists who were grasping for empirical data. The nature of the references by specialist consultants indicates that the putative data on Ethiopia's deforestation did not consist of new empirical research but conjecture circulating orally and accepted as fact among the expatriate scientists who formed Ethiopia's early core of development planners.[5]

The citation of the 40 percent historical forest cover figure and concern about the rapid pace of Ethiopia's deforestation over the twentieth century has not been merely an aberration of foreign consultants. Hoben has noted that such development narratives often take root within developing nations themselves, lending them an aura of authenticity. Early speculation by expatriate experts and teachers passed into the accepted knowledge of the first generation of Ethiopian biologists, foresters, and geographers. By the early 1980s the "facts" of Ethiopia's historical forest cover were standard boilerplate for environmental reports and appeals for development aid. The United Nations Development Program and the World Bank, for example, had picked up the figure and published it officially, citing it in a joint 1983 report on Ethiopia's energy sector.[6] By the mid-1980s Ethiopian official documentation had also begun citing the figure. In 1985 the Ethiopian Relief and Rehabilitation Commission published an official account of the causes of the Ethiopian famine, arguing that deforestation had caused drought and, indirectly, famine. That publication cited 44 percent forest cover in 1885, 16 percent in 1950, and 4 percent by 1985.[7]

In later years Ethiopia's deforestation narrative had leaped to a new generation and embedded itself into the country's national self-image. In 1994, for example, a young Ethiopian forester on the staff of Alemaya University offered the 40 percent dogma as evidence in a paper presented to a university conference on land tenure sponsored by Addis Ababa University and the Ford Foundation. Most recently, the deforestation trope appeared in papers delivered by three noted Ethiopian policy makers at a national conference on population and natural resources in Addis Ababa and in an address by the Vice Minister of Agriculture to a development workshop.[8]

No doubt the statistics of Ethiopia's twentieth-century deforestation have been repeated at meetings, in policy briefings, in hotel lobbies frequented by expatriate experts, and in university lectures on several continents, all without a foundation in evidence from Ethiopia's landscapes themselves. More recently, reliable land use surveys based on Landsat images have appeared in the public domain and are in a position to quantify the current state of forest resources. None of these recent surveys, of course, have any data for 1900 or 1950, the historical data that is fundamental to calculate the rate of forest loss.[9]

There are three important elements of Ethiopia's forest degradation narrative as described by Allan Hoben's definition offered in Chapter One. First, it asserts a historical baseline for Ethiopia's national forest resources at a particular point in time (1900 or 1950) for which no such systematic survey data are available. There is an implicit argument that ecological resources prior to the benchmark year were in a better state. Second, there is a specific argument made about degradation: that it is cumulative and has its origins in human mismanagement or at least population growth. Finally, there is an apocalyptic forecast: the expectation of a trajectory to disaster when Ethiopia's forests apparently disappear altogether, resulting in chronic drought (as Gore has it) or other calamities. This chapter will address each of these points.

Modern narratives of resource exhaustion in Ethiopia have a powerful resonance among observers of highland landscapes of the late twentieth century simply because the Ethiopian highlands in fact show an extraordinarily heavy imprint of human action accumulated over at least two and a half millennia. Deep soils, dedicated farmers, and a distinctive African technology—the single-tine scratch plow called the *maresha*—allowed early agricultural innovation in cereal crops and the evolution of a remarkable arsenal of agronomic strategies that highland farmers have deployed within the highlands' kaleidoscope of local microecological settings. Indeed, the general theme of systematic exhaustion within the deforestation narrative fits a basically accurate analysis of the classic plow-based farming system as extensive rather than intensive. The labor efficiency of the oxplow, in fact, has historically fostered a short fallow cultivation of cereal crops that required clearing land of trees and scrub to allow yoked oxen to maneuver and the plowman to swing his plow beam freely on turns. Fencing, hedgerows, or windbreaks of any sort were exceedingly uncommon, most fields being marked by a rock or the memory of the cultivators. Wood lots as land use practice were nonexistent until the late twentieth century. That an expanding frontier of plow agriculture over the course of time changed the landscape and affected the nature of its vegetative cover is not in dispute.[10] Deforestation has historically been part and parcel of highland agriculture. What is at issue, however, is the rate and meaning of that change in historical context.

There is consistent historical evidence that over the past millennium Ethiopia's highland farming system had evolved a full repertoire of practices of intensive cultivation and resource conservation that farmers could bring to bear as needed in particular places and at particular times. These included terracing, irrigation, contour plowing, elaborate crop rotation schemes, and fertilizing of household garden plots with both manure and hearth ashes. In some settings oxplow cultivation has accommodated perennial crops and root crops alongside annually cultivated plots.[11]

In the highland oxplow rural economy wood constituted an important but not a critical off-farm resource, gathered by men as needed for the building of houses or granaries and to serve as plow parts. By far the largest potential wood requirement, however, was for daily household fuel consumption: baking, preparation of stews or boiled pulses, or brewing. Though women and men might conceivably cut branches or entire trees for this purpose, women also have employed a range of options that are less labor intensive than cutting live wood, including the use of dried dung, dead wood, sorghum stalks or wheat straw. Though highland Ethiopian women in the late twentieth century indeed have had to expand their foraging for fuel materials, deforestation is not the only factor increasing the scarcity of fuel supply.

Population pressure on open pasture has also meant smaller livestock herds that produce less manure per household and increased competition for straw as livestock forage.

Contemporary observation of rural life in Ethiopia thus seemingly provides circumstantial evidence in support of a story of long-term land use practices that have degraded the landscape and drastically reduced the highlands' vegetative cover. Modern conservationists often support their degradation perspective by citing a long-standing argument that Ethiopia's cities and towns have historically consumed their own ambient forest cover, thus accelerating deforestation.[12] But does the historical evidence support these accounts and the general deforestation narrative?

The following cases review the evidence from two areas of Ethiopia, reflecting two separate zones of Ethiopian land use history and vegetative cover for the nineteenth century, a period when the degradation narrative implies forest and vegetative cover were more extensive and stable prior to the precipitous decline of the twentieth century (see Figure 5.1). In Case 1, I examine the historical record of Ankober, a region north of Addis Ababa that lies on the edge of the highland's eastern escarpment. As a zone that contained the nineteenth-century capital of the kingdom of Shawa, Ankober offers both dense historical description by travelers and an example of a dry evergreen montane/grassland typical of the central highlands. Ankober offers another dimension as well since historians and geographers have often argued, or assumed, that depletion of its fuel supplies motivated the movement of the royal capital from Ankober in the 1840s.[13] The treeless plain over which the modern road to Ankober passes has reinforced the premature conclusion of modern observers that Ankober's settlement denuded the hillsides over the course of less than a century of intensive occupation. The nineteenth-century evidence, however, indicates that the condition is not new; Ankober and much of the central highland landscape has been devoid of trees and wood fuel for at least one hundred fifty years and almost certainly much longer.

In Case 2, I examine the history of Gera, a region within the southwest's broadleaf forest, amidst the area of highest concentration of Ethiopia's current forest resources. Here the historical evidence refutes the deforestation narrative's view of a linear, cumulative process of loss of vegetative cover. The nineteenth century was a period of significant dynamism that reflected politics, local social institutions, and the impact of oxplow agriculture on broadleaf forest ecology.

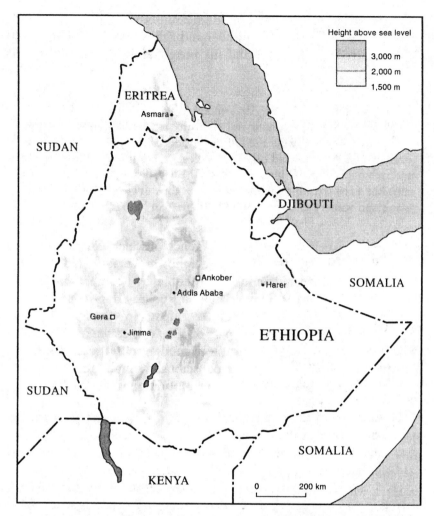

Figure 5.1 Map of Ethiopia.

While these two cases cannot present a comprehensive survey of the entire highland landscape, they do, I believe, establish the fallacy of ubiquitous Malthusian narratives that claim a historical baseline.

Case I
Alpine Groves or Denuded Hills:
Ankober's "Forests" in the Nineteenth Century

Mid-nineteenth century observers of Ethiopia's highland land-scapes described what they saw by using a different narrative style and set of metaphors than their modern counterparts. Major W. Cornwallis Harris's first view of Ankober and Ethiopia's central highlands in 1840 stirred his emotion and his imagination. He described the sight as

> a magnificent view of the Abyssinian Alps. . . . Hill rose above hill, clothed in the most luxuriant and vigorous vegetation; mountain towered above mountain; and the hill-clad peaks of the most remote range stretched far into the cold blue sky. Villages' dark groves of evergreens and rich fields of every hue, chequered the broad valley; and the setting sun shot a last stream of golden light over the mingled beauties of wild woodland scenery and the labours of Christian husbandmen.

Three years before Harris, the French traveler Charles Rochet d'Héricourt had made the Alpine metaphor more explicit, describing his first view of the highlands at Ankober as "comparable to the most beautiful sights of Switzerland."[14] Rochet d'Héricourt's Swiss imagery and Harris's references to "groves of evergreens" and "woodland scenery" were thus part of a style of travel narrative that highlighted the subjectivity that more-modern narratives attempt to suppress, and their descriptions consciously sought to confirm to European readers highland Ethiopia's reputation in Europe as an exotic but familiar place.

These images of the highlands, however, were not absolute but relative. Those European travelers who encountered the central highlands for the first time near Ankober had had to approach the highlands after several days of trek through a hot lowland scrub at sea level. Moreover, struggling across the terrain from the Red Sea coast also meant crossing unfamiliar ethnic and religious space, from the bleak Islamic coastal plain to an almost temperate Christian highland. Choices of sensuous images of a blue, green, and cool habitat of Christian husbandmen contrasted with the hot, flat, and dangerous terrain occupied by Muslim pastoralists in the lowlands east and below Ankober.

Though some nineteenth-century travel accounts of Ankober published in geographical journals, as exotic travel literature or as private correspondence, are hyperbolic in their prose, they are nonetheless an excellent means of reconstructing the historical landscape and determining the state of the vegetative cover for the period 1840–86 when Ankober served as one of Shawa's several royal towns prior to Addis

Figure 5.2 Ankober terraces (from a photograph), 1888. From J. M. Bernatz, *Scenes in Ethiopia, Vol. 2: The Highlands of Shawa* (Munich and London, 1852).

Ababa's foundation as imperial capital. While the narrative themes of some accounts play to their European audience's expectations of Ethiopia as an Alpine Christian Shangri-La, many travelers were keen and observant naturalists and experienced geographers who offered quite precise descriptions of landscapes and ecology. Taken as a whole, these accounts offer thick, contemporaneous descriptions of landscapes that local oral tradition and living memory cannot provide. Early photography and landscape engravings also provide an important record of the state of ecology and land use.

The trickle of observers who visited the Ankober region of Ethiopia's central highlands in the 1830s and 1840s were attracted to a successful and expanding polity built upon an agricultural base that strongly resembled highland farming systems of the late twentieth century. Northern Shawa, the region of Ankober, was the home of the Amhara people who in the mid–nineteenth century felt themselves to be heirs to the restored Solomonic dynasty and had recently pushed out Oromo residents and reestablished (in their own view) a political kingdom on their traditional "homeland." Thus, the Ankober region, though bitterly contested at times, was a zone of habitation long oc-

Figure 5.3   Interior of Ankober town, c. 1840. From J. M. Bernatz, *Scenes in Ethiopia, Vol. 2: The Highlands of Shawa* (Munich and London, 1852).

cupied by a highland oxplow rural economy. Ecologically, Ankober consisted of a typical highland ecozone, hemmed in on the east by lowland savanna acacia scrub.

Descriptions of the highland landscape around Ankober in the nineteenth century bear little resemblance to the forested zones envisioned by twentieth-century degradation narratives. If some parts of early narratives of Ankober were enraptured by the idyllic beauty of Ankober's immediate environment above the lowland wastes, eyewitnesses were nevertheless shocked and often appalled at the lack of forest and wood fuel on the central highlands of Shawa as a whole. Those who began their accounts with optimistic Alpine images, generally offered more dyspeptic descriptions of highland vegetative cover in the body of their published work. In 1838 Rochet d'Héricourt in a more prosaic mood noted that forests in his Alpine Shawa were actually "quite rare," while Harris in 1840 referred to areas on the highlands west of Ankober as "long naked sweeping plains . . . destitute of wood." Traveling for several days on Shawa's southern frontier contiguous with Ankober, Harris recalled that "not a single bush or tree was visible during the long ride." Dr. Charles

Figure 5.4 View between Entoto and Ankober, 1888. From Jules Borelli, *Éthiopie méridionale: Journal de mon voyage aux pays Amhara, Oromo et Sidama, septembre 1885 à novembre 1888* (Paris, 1890).

Johnston, writing about the same landscape in the same period described the terrain as "bleak, moor-like."[15] In 1840 Douglas Graham, a sober and experienced agronomist with experience in India, compiled a detailed set of agricultural and land use observations and ventured that "from a careful observation during many journeys in every direction, there are few forests or wastes, excepting the impractical for pasture or cultivation."[16]

Some sixty years later (e.g., 1900), Augustus Wylde, an experienced English traveler arriving in Shawa via Sudan, looked southeast from the edge of the Adabai River gorge a day's march northwest of Ankober district and reported seeing no trees at all on the horizon. Jules Borelli's 1888 engraving based on a photograph shows the area between Ankober and Entoto (present day Addis Ababa) as the barren plain mentioned by earlier travelers (see Figure 5.4). Obviously, the blue-green forest familiar to visitors to Addis Ababa in the 1970s was indeed a twentieth-century phenomenon, consisting exclusively of eucalyptus planted there after 1902.

While some deforestation was doubtless a product of the oxplow's long-term occupation of the highlands, the ecological setting itself limited the amount of tree cover it could sustain. Christopher Clapham has cleverly pointed out that the species of endemic birds that evolved on the highlands are ground-dwelling species and not forest types, indicating a long period of open savanna habitat. Fire, as in other savanna/prairie zones, also favored grasses over trees. Native tree species tended to collect in well-drained valleys on niches of alluvial soils or near natural fire breaks, rather than as forest cover per se. Many areas, such as Shawa's extensive vertisol plains suffered annual waterlogging that prevented tree growth. Manoel de Almeida in the seventeenth century reported as early as the seventeenth century that "generally speaking, there is not much woodland in Ethiopia."[17]

While there were few trees on the highlands, there appears not to have been a nineteenth-century deforestation/fuelwood crisis. Ankober town itself and the royal compound in the 1840s dealt with the lack of wood and fuel resources by managing its own fuel supply and domestic forest. As today, the town itself may have had more vegetative cover than the surrounding countryside precisely *because* of the concentration of population there. It was, more than likely, the town's junipers and ornamental banana trees visible from the lowland approaches that inspired travelers' Alpine nostalgia.

It seems likely that trees were then, as in Ethiopia today, more a sign of settled urban life than of bucolic nature. Ankober's domestic and church compounds protected juniper as shade trees; huge wild fig trees denoted a central marketplace; and cultivated banana plants appeared as ornaments in domestic compounds, their leaves serving as wrapping for bread.[18] Those trees and their protection may also have had some value as symbols of authority. When Emperor Tewodros's troops overran Ankober in 1859–60 on a punitive expedition, his occupying troops cut down about half of the royal town's trees before burning the town. Trees in this case served more as markers of royal authority than as fuel, their destruction was a symbolic act, not evidence of environmental insensitivity.[19]

Ankober's wood fuel resources were, it seems, never its own junipers, wild fig, or podocarpus but forested niches in surrounding microecological zones. Probably to minimize effects of wood collection on the immediate hinterland, the king maintained a group of three hundred slaves specialized as cutters of wood who supplied the royal palaces at Ankober, Dabra Berhan, and Angolela with juniper and wild olive logs cut from royal forests at the heads of the steep river valleys below the escarpment. The king allowed each slave three days to gather and deliver each load, an indication of the remoteness of the royal forests even at the apogee of Ankober's power.[20]

The most extensive of the escarpment forests was the Feqre Gemb, a primary forest located north of Ankober town, adjacent to a former royal pasture and still visible to the left as one approaches Ankober by the modern road. In 1879 the Italian naturalist geographer Orazio Antinori reported over twenty species of trees, twenty-two species of shrub, and abundant game. The Feqre Gemb is a superb example of primary climax forest, but its setting suggests it was not a remnant forest but one adapted to a specific environmental niche. That the forest has survived through the twentieth century is attributable to its inaccessibility in a steep valley and, at least in the nineteenth century, its protection by royal decree and the presence of a nearby monastery.[21]

As observers indicate, wood, then as now, was a rare, protected commodity, and peasant households had to rely on alternative household fuels, especially the use of dung collected from domestic livestock. Late-twentieth-century environmentalists have often cited the use of dung as an indicator of a deforestation crisis. Moreover, the degradation narrative argues that the use of manure to preserve soil fertility has declined in recent years because of its increasing use as fuel. This argument is contrary to the facts of the Ankober evidence. In the entire 1840–90 period the historical evidence indicates that manuring was limited to a few selected plots and never applied to fields in a systematic fashion. Graham, for example, noted in 1841 that difficulties of transport, for example, no wheeled carriages, meant that manure was never used in fields, save that deposited by livestock tethered on harvested plots to feed on stubble (a form of intensification).

Far from dramatic changes in fuelwood/dung uses on the central highlands in the late twentieth century, there has been little change visible in the use of dung over the past two centuries. Augustus Wylde in traveling around 1900 in northern Shawa noted the exclusive use of dung for fuel.[22] In 1925 Rosita Forbes, a traveler sensitive to women's work obligations confirmed these observations, describing an absolute lack of wood fuel on her route between Ankober and Addis Ababa. She noted, as had Wylde a quarter century earlier, that cattle dung was the dominant household fuel. Eighty years earlier Graham in 1840 had already complained about the stench around

Figure 5.5 Faggot sellers, northern Shawa (from a photograph), c. 1880. From Jules Borelli, *Éthiopie méridionale: Journal de mon voyage aux pays Amhara. Oromo et Sidama, septembre 1885 à novembre 1888* (Paris, 1890).

Shawan houses deriving from the farmers' habit of stacking dried dung fuel close by their doorways (see Figure 5.4).[23] In other words, the carefully collected and dried piles of dung around northern Shawa homesteads that are visible to vehicles traveling on the main tarmac road in recent decades were as common a sight to nineteenth-century travelers on muleback. Women's elaborate typology of dung fuel described by Helen Pankhurst in her recent field work in the central highlands displayed a repertoire of on-farm resource use that women had evolved over generations of local practice, not a recent response to environmental degradation.[24]

In short, Ankober's nineteenth-century landscape was not a heavily forested highland but one that looked then remarkably as it does today, minus the ubiquitous eucalyptus: open highland plateau, a few primary forests in inaccessible zones, and agricultural fields whose color and texture change with the season and the rotation of annual crops. Historical Ankober and its environs apparently had no more tree cover than it does today, though, ironically, the royal town may have had the region's most visible tree cover in the 1840s when it had a higher population than it would have in 1900 or 1999.[25]

Figure 5.6 Rural Shawan house, c. 1888. (Note dung pile at left of image.) From Jules Borelli, *Éthiopie méridionale: Journal de mon voyage aux pays Amhara, Oromo et Sidama, septembre 1885 à novembre 1888* (Paris, 1890).

Case 2:
Plow, Forest, and Landscape History in the Broadleaf Forest: Gera 1850–1940

Though Ethiopia's oxplow agriculture had its origins in the relatively open savanna woodland environment of the northern highlands, it also spread south and west across the Nile and onto an entirely different cultural and physical landscape. In the Gibe River basin in the early nineteenth century a group of small, well-organized Oromo kingdoms emerged in a broadleaf forest zone to take advantage of a burgeoning export trade in extractive trade goods—slaves, civet musk, and aromatic herbs and spices.

The dense broadleaf montane forest that historically covered the highlands of Gera in southwest Ethiopia contained dense stands of hardwoods exceeding 35 meters in height and, compared to northern open and often treeless plateaus, was a diverse forest ecosystem that supported a staggering variety of natural products. In the primary canopy, Emilio Conforti in 1939 counted thirty-five species of "tall trunked" trees. In 1991 my afternoon's trek up the forested slopes still revealed similar and significant biodiversity: thirty-three locally recognized species of trees, ground cover shrubbery, root crops, and broadleaf plants, all with specific uses in the local economy. It was also within this moist, shaded curtain of vegetation that Ethiopia's rich varieties of coffee *arabica* evolved in a wild state.[26]

Mid-1980s survey data indicate that the broadleaf forests of the southwest made up 65 percent of Ethiopia's total forest resources.[27] The history of that forest cover, however, remains poorly understood. Over the course of the nineteenth and early twentieth centuries the interaction of the oxplow system, population, and the forest environment shaped Ethiopia's southwestern forest in complex ways missed by contemporary arguments about deforestation.

Gera's forest ecology contrasts starkly with the northern cradle of oxplow agriculture as represented by Ankober. An important feature of change within the southwest's human/forest landscape was Gera's substratum of forest agriculture that predated plow cultivation and included food crops adapted to a shaded forest environment and horticultural forms of cultivation. This tradition included a bundle of domesticated root crops well adapted to forest soils: taro, yam, and *Oromo dinich* (*coleus edule*). The repertoire also included *ensete* (*Ensete ventricosa*), or false banana, a crop widely considered to have predated annual crops in several highland cultivation systems.

Our images of historical land use and forest ecology in Gera come from local oral narratives, recent observations of existing forests, and a succession of Italian travelers who have provided vivid and detailed observations of the forest landscape in detailed snapshots in the 1850s, 1880s, and again in 1928. Their narration of the forest setting included in each case a detailed prior knowledge of the central highlands and an intense curiosity about Gera's forest context. The forest appears in their narratives as an exotic element, but also a living part of the landscape. Each traveler described successive changes in the ecology and its interaction with its human population, giving us a sustained longitudinal portrait of the state of the forest.

Local narratives offer a surprisingly negative view of the forest's human ecology. In Gera local legend describes a malediction issued by a deposed king, Abba Bosso, against his own disloyal subjects who had usurped his throne, imprisoned and blindfolded him. He cursed his people and their land thus: "As you gave me perpetual darkness, let your country be swallowed up by forest. If you cut a single tree let it be multiplied by ten. May you not see the light, only darkness."[28]

Abba Bosso's curse reveals an attitude about trees and forest ecology as a bane on human happiness but offers no description of a historical landscape. For that we must rely on written accounts from Italian observers.

The first of these accounts belonged to Guglielmo Massaja, a Cappuchin missionary who established two mission stations in Gera in 1859 amidst a thriving agricultural landscape carved out of the vibrant forest ecology. Two decades later, Antonio Cecchi, a geographer, spent a year at the same sites, providing an even richer description of a diverse system of annual cropping mixed with forest crops and the local social/political context. In 1928, that is, almost fifty years later, Enrico Cerulli, a brilliant young linguist/ethnographer, infatuated with the new Italian nationalism, set himself the task of retracing the steps and rekindling the national glory of his compatriots' accomplishments. He described with great narrative force the resurgence of the climax forest canopy that followed a massive depopulation. In 1990 and again in 1991, I conducted field work in Gera, retracing the steps of these earlier observers and merging their accounts of landscape change with local oral tradition.[29]

Travelers to the southwestern forest viewed it quite differently from the late-twentieth-century narratives that portray its vegetative force as a passive victim. For early travelers the forest was an active, growing presence, beautiful but threatening. Massaja saw his mission of religious enlightenment to be coterminous with establishing open, settled agricultural landscapes to support his followers. From his perspective, the forest was a threat to that mission. Cecchi was ambivalent, describing the forest sometimes as "majestic . . . colossal" and at other times as "dark . . . menacing." For Cerulli, the forest was an everpresent force bent on obliterating traces of Italian achievements. Its resurgence into a climax canopy represented nature's reclaiming of its own.

These images of the nineteenth-century forest derive from a ground-level perspective—not an aerial view but a traveler's muleback view that moved through forest to clearings, to the royal compound, along well-marked routes of travel created by movements of Gera's own population of farmers and traders. Travelers first entered Gera along a caravan route from the east. Political conflicts between these small states had already shaped the forest; differences in economic structure and land use set these states apart from one another in ways that were written on the landscape itself. What appeared on Gera's borders as natural forest was, in fact, a political creation. The frontier of Gera with its hostile neighbors consisted of socially vacant spaces designated by political conflict as neutral zones with no settlement. Cecchi described the frontier zone he crossed as "immense and gloomy forest, rich in time-honored trees with colossal trunks. Among them were podocarpus, euphorbia, cussi, sycamore, wild olive, acacia, and gardenia." Passing through the "majesty of the superb forest" toward central Gera, suddenly, he tells us, the dense vegetation broke: "In the valleys and along the slopes of the hills are the plowed fields with the scattered frequency of the thatched huts and villages, posted in the most pleasant positions." Further east, approaching Gera's royal capital, Cecchi described extensive areas covered not at all by forest canopy: "The end of this valley presents an extended plain, little cultivated although the soil was very fertile and rich in water which for lack of slope here and there was waterlogged. The dry part forms a luxurious meadowland where beautiful cattle and sheep grazed."[30]

Narrative descriptions from 1859 and 1880 testify to a forest cover, not in a primeval state, but already in retreat in the face of human settlement, especially the needs of annually cropped cereals. Cecchi describes homesteads that were set back from these open plains, either isolated or collected in small hamlets of ten or fifteen families. Gera families constructed their houses, "with a certain air of elegance" of bamboo, ensete leaves, and teff straw, revealing the products of their mixed, oxplow farm economy.[31] At the capital of Gera, at Chala (or Ciala), the royal compound was set—as is the home today of the royal family's descendants—on one of a chain of hills at the western edge of the bottomland meadow. Massaja recalls it in 1859 as "dressed with plants of citrus, coffee, bananas." From Cecchi's description, it is clear that the oxplow complex's annual cropping system was well in place in Gera by 1880. He described the principal products of the Gibe states as "different qualities of teff, maize, sorghum, eleusine, wheat, barley, peas, fava beans, and haricot beans."[32] These food crops reflect the coexistence of annual crop agronomy (seed broadcasting) with horticultural traditions of vegetative propagation (root crops and rhizome plantings), which suggests that forest clearance for cereal crops with oxplow was already well established by mid-century. The illustration published in both Cecchi's two volumes and Massaja's account in the 1880s (see Figure 5.7) shows the royal compound on an open

Figure 5.7   Gera's royal residence, c. 1860. From Antonio Cecchi, *Da Zeila alle Frontiere del Caffa.* 2 vols. (Rome, 1886).

rock outcropping surrounded by ensete and lemon trees, a distinctly human-shaped landscape.[33]

The capital itself was a substantial settlement, consisting of the royal compound, animal enclosures, and quarters for 3,000 of the king's slaves who prepared food for the royal household and cultivated extensive royal fields of annual crops.[34] Local memory has exaggerated the forest's retreat but makes the point emphatically. Abba Dura, an eighty-year-old native of the area, recalled for me what he had heard as a child of the nineteenth-century open landscape around his home a stone's throw from the old capital: "There was no forest. People had to use dung as fuel. The whole area was covered with houses. There was no forest; there was nothing to use for fuel. There was no wood!"[35]

The landscape of the district of Afallo to the south and west of the capital toward the Gojeb River valley tells a similar story. Afallo was, at least in the nineteenth century, the densest settled region of Gera, even as today it is the most heavily forested. Yet in 1858 Massaja described the people as herders and farmers whose homesteads were scattered across beautiful and "flowing" cultivation. At Afallo, Massaja established a Capuchin mission station, planting the open grounds

Figure 5.8   Afallo mission, c. 1860. From Antonio Cecchi, *Da Zeila alle Frontiere del Caffa*. 2 vols. (Rome, 1886).

with ensete, coffee, lemon, teff, and "every sort of indigenous cereal."[36]

Twenty years later, Cecchi retraced Massaja's path to the site of the mission station. Like Massaja, he was impressed with its openness and its human-shaped landscape: "The country is the most beautiful and picturesque, and most populated of the kingdom. Everywhere thatched huts separate or together in groups in the midst of ensete plantations, lemon trees, coffee,—cabbage trees [kale] a special variety of potato, onions, garlic."[37]

The landscape engraving of the mission at Afallo published with Cecchi's book in 1884 depicts a bucolic but cultivated garden setting: open ground surrounding the mission dwellings, ensete interspersed with fruit trees. One can imagine at a distance the mission's fields of cereals described by Massaja, cultivated each season at that point for over twenty years. The image of an open landscape is congruent with both observers' textual accounts of the site as densely settled and extensively cultivated in annual crops and ensete. Abba Fogi, a resident of the area who spoke to me in 1991 recalled descriptions from his grandfather's time, that is, the nineteenth century, that the district was "a field without any trace of forest."[38]

Figure 5.9 Afallo mission site, 1991. Photo by author.

Given the density and resilience of the forest cover, how did Gera's agricultural base achieve its nineteenth-century dominance? With the existing level of technology of agriculture and implements for cutting trees and vegetation, only a rapid rise of population and labor supply could account for the massive early-nineteenth-century forest clearing implied in the open lands at Afallo and along the central zone around the royal compound that, except for marshy bottomlands, had previously supported forest.

Dating this transformation is difficult, but it is reasonable to assume that the foundations of the kingdom rested on the maturation of cereal agriculture as a means of accumulating, concentrating, and storing food, a process far more difficult with forest food products. The expansion of the oxplow complex in both Gera and the southwest forest region as a whole thus took place simultaneously with the application of concentrated labor power on the forest cover and the formation of small-scale kingdoms. Once cleared, fallowed forest plots with their characteristically shallow nutrient base needed three years' fallow every five years, and new plots had to be cleared from the forest. Once cleared of virgin cover, however, forest plots required significantly less labor to bring into production again.[39] Over time, the labor invested in clearing forest plots reached a plateau and leveled off to a point of equilibrium where retilling fallowed plots was the norm rather than clearing virgin forest.

The historical evidence of changes in Gera's forest challenges images of a primeval forest pushed back steadily in a linear progression of human effort. Far from a progressive front of cumulative agricultural expansion, the evidence supports an episodic, conjunctural process whereby open, cultivated land dominated the landscape wherever labor allowed, and a resilient forest cover encroached once again whenever human labor diminished. What we know of historical property systems and land tenure in Gera suggests that Gera attracted populations from surrounding zones in its early settlement history. Fertility rates were far less critical to Gera's historical labor supply than the rapid absorption of new immigrants. Local social practice encouraged such movement. Migrants and refugees did not face a rigid ethnic definition of land ownership, and Oromo property law transferred land and property to the eldest son but provided income to younger sons and dowries to daughters, encouraging and capitalizing on clearing new plots in the forest itself. Oromo lineages also apparently allowed households without issue to adopt fictively new members and endow them with the property rights of children.[40]

There were other forms of labor as well. Cecchi in 1880 estimated Gera's population at between 15,000–16,000, an estimate that appears not to have included the huge population of slaves (3,000 in the royal capital alone) he reported elsewhere in his account.[41] Slaves' most important long-term role was as a captive class of agricultural labor. Free farm households, many of which may have consisted of migrants, also worked royal lands, but the householders served as tenants who probably bore responsibility for clearing new forest fields as part of their tenancy.[42]

In the period following Cecchi's 1881 departure from Gera, regional politics, population, and forest ecology intertwined to produce a new forest landscape by the turn of the century. In 1881 Menilek's imperial Shawan armies invaded Gera, driving the Queen Mother and royal family into exile. From contemporary accounts and oral evidence, there appears to have been a massive exodus of population. Gera's free population fled south while the invading armies shipped Gera's own slave population to the north as booty. Frank de Halpert, head of Ethiopia's antislavery commission in 1932, estimated that the population of the region "probably decreased by three-quarters," a figure confirmed in local traditions. In fairly short order, the absence of labor permitted the forest to recapture what it had yielded to the plow, machete, and short-handled ax only a few decades before. By the late 1920s virtually all of the open cultivation had reverted to climax forest. In 1900—the deforestation narrative's benchmark year—certainly the forest had enjoyed almost two decades of regrowth to recapture territory earlier lost to the plow.

An enduring theme of agriculture in Gera and in the narrative historical accounts has been the subtle, dialectical interaction between human settlement and the forest cover. The forest itself was not primeval and unyielding; when I first saw its vegetation in 1990, it clearly bore the marks of several historical layers of human action in the form of recently fallowed clearings, secondary growth of specialized quick maturing sun-loving trees later dominated by the resurgence of tall species that blocked sunlight from reaching the forest floor. Local residents can recount in great detail the progression of growth of forest species. Within the forest on a fallowed plot or abandoned homestead site, the first stages of forest, soft wood saplings, appear within a year along with bushes of a wild clovelike aromatic, amomum (ginger root), "monkey peas," baboon's cabbage, and flowering *tufo* whose nectar makes up one type of the region's superb honey. Within four years mimosa and hardwood saplings provide sufficient shade for spontaneous growth of coffee seedlings. The broadleaf forest's tallest trees also grow surprisingly fast, creating a canopy 20 meters or above within ten to fifteen years. The "climax" forest in Gera develops within forty to fifty years and includes hardwood trees like the Sombo (*Ekebergia capensis*), which reaches 45 meters, with a trunk diameter of 2 meters. The fifty-year climax forest thus may include thirty to forty species of trees, shrubs, and large plants and a high canopy that might reach 50 meters.[43]

Testimony to the forest's regrowth in the twentieth century and the effects of the flight of human labor from Gera comes from Enrico Cerulli who revisited Gera in 1928 to retrace the steps of Cecchi and to locate the site of Massaja's mission in the interests of both ethnological science and nationalism. Cerulli arrived at the Gera frontier, he noted, a month and a day from the forty-ninth anniversary of Cecchi's entry, near the fifty years needed for a new climax forest growth cycle. After four hours' trek through the frontier forest, Cerulli described the effects of forty years of secondary forest growth on Gera's once thickly settled central highland:

It was possible to discern only—very indirect alas!—traces of human life. . . . Those that were one time the names of villages, were now passed and indicated simply by some tracks in the forest; in some years these will be entirely forgotten. At Filo Cecchi had found a village and from this rose at that time Ciala [Chala, the royal capital]; now there is nothing, not even a hut and the forest has covered all the hills, not leaving anything to see.[44]

The 1939 Italian *Guida dell'Africa Orientale Africana* travel guide described the site of the former flourishing capital for potential visitors as "a few huts menaced by the forest."[45]

Cerulli continued his trek along the old route past the old capital southwest to the site of the mission station that in the mid–nineteenth century both Massaja and Cecchi had called Gera's most densely settled region. He found the open, cultivated plain depicted in Cecchi's 1880s engraving and narrative entirely reclaimed by the forest. The mission station itself was covered in a dense secondary growth of hardwood trees and ground cover. Cerulli's described the result of the aftermath of depopulation and almost a half century of regrowth:

We have not seen even a village, we have not seen even one man. The forest is most thick and intricate. . . . The mission was clear forty years ago; and with this equatorial climate and with this humidity an abandoned village situated at the margins of the forest, disappears, victim to the forest, in still less time.[46]

The history of Gera's forest cover in the nineteenth century was not a progressive movement from nature to human settlement but a complex tale of shifting vegetative landscapes dominated in turns by plows and trees. In 1991 when I visited Afallo, the residents helped me locate the old mission site. The area remained enveloped by a dense climax forest canopy, remembered as open cultivated land only by the oldest residents as a story two generations past.[47] Local memory also contains its own narrative on the meaning of Gera's fecund forest ecology. The curse of Abba Bosso quoted earlier suggests local views of forest as the absence of vision and light, its growth a threat or a curse. This view contrasts markedly with a development narrative that equates forest with idealized nature.

CONCLUSION

Laid against the narrative of deforestation posited by Albert Gore and the conventional wisdom of development, neither historical case examined here bears out the premise of a heavily forested nineteenth-century Ethiopian landscape. Forest cover for Ankober and the central highlands was already an ancient memory by the time the first records of its landscape appear. The evidence from Ankober points toward the management of scarcity, an accumulated arsenal of techniques evident at the farm-level and state-level management of key assets like pasture, forest, and water. Then as now, the presence of trees on the central highlands meant the presence of the human hand and concentrated population rather than a state of nature.

The 1984 famine in northern Ethiopia created a sense of urgency among international donors who wished to provide immediate relief and also provide for development efforts to reduce the probability of a recurrence. With support from the EEC, the United States, and many other donors, the Ministry of Agriculture began a massive program of reforestation to reduce erosion, create local fuel resources, and re-create a "lost" forested landscape in Ethiopia's northern highlands. The reforestation program planted over 300,000 hectares of tree seedlings on the highlands between 1984 and 1990. By 1992 most of those seedlings had died from livestock damage, fuel foragers, or ill-suited placement. Though farmers received grain and cooking oils for their labors in planting millions of seedlings per year, the attempt to re-create a putative forested landscape failed.[48]

With the best of intentions both the Ethiopian government and international donors had rushed to embrace the narrative of Ethiopia's forested past. Ironically, it was the Swiss who had conducted much of the highland reclamation research used as a model for the tree-planting program. Though the Swiss may not have known, they may have fallen into the same false Alpine analogy that afflicted some nineteenth-century travelers, that is, they imagined a Swiss landscape for the central highlands, one that was neither supportable ecologically nor historically accurate.[49]

Gera's forest canopy in 1900, impressive as it might have been, was in a state of vigorous regrowth after a long period of steady retreat in the face of the plow, machete, and a growing population. Cerulli's description of a climax forest, he knew, was not a static state but a new assertion of vegetative expansion. Ethiopia's southwest forests were indeed changing, as they had throughout the nineteenth century, but in directions that were neither linear nor cumulative.

Neither of these stories, in fact, offers a smooth linear tale, nor can they quantify Ethiopia's forest cover at any given time. Rather, they amply demonstrate that environmental history is not usually a progressive, cumulative march forward in time toward current conditions. The history of the environment of a given setting, society, or epoch is a combination of disparate factors of physical forces, technology, context of political economy, and population. It is, if not serendipitous, certainly conjunctural.

NOTES

1. Albert Gore, *Earth in the Balance: Ecology and the Human Spirit* (Boston, 1992), 107.

2. H. P. Huffnagel, *Agriculture in Ethiopia* (Rome, 1961), 395, 405–6. Huffnagel estimated a 1.4 percent annual loss in forest cover.

3. F. von Breitenbach, "National Forestry Development Planning: A Feasibility and Priority Study on the Example of Ethiopia," *Ethiopian Forestry Review* 3 (1962): 43.

4. W. E. M. Logan, *An Introduction to the Forests of Central and Southern Ethiopia* (Oxford, 1946), 23–27.

5. There is no evidence that the 1954 U.S. Army aerial photography project issued any data on forest cover. The first national land use surveys by the Imperial Ethiopian Government appeared in the late 1960s but contained no survey data on forest cover.

6. United Nations Development Program and World Bank, *Ethiopia: Issues and Options in the Energy Sector*, Report No. 4741-ET, p. ii.

7. Relief and Rehabilitation Commission, *Combatting the Effects of Cyclical Drought in Ethiopia* (Addis Ababa, 1985), 34–35, 155–6. Addis Tiruneh, "Gender Issues in Agroforestry," *Proceedings of the Second Workshop of the Land Tenure Project* (Trondheim, 1994), 1. In the interests of self-expiation, I must also admit to have found the reference to the 40 percent figure in my own earlier work to indicate wider effects of local processes indicated in my fieldwork. See James C. McCann, *From Poverty to Famine in Northeast Ethiopia* (Philadelphia, 1987), 34. Arguments about deforestation in modern Ethiopia are not without foundation in lived experience. During field interviews, older residents of the highlands have told me they recall forested areas that are now intensively cultivated under annual crops. Most such stories relate to the 1940s or 1950s.

8. Association of Ethiopian Geographers, "First Annual Conference of the Association of Ethiopian Geographers: Population, Sustainable Use of Natural Resources and Development in Ethiopia: Program and Abstracts," May 31, 1996, pp. 8, 15, 21. "Forests Go from 40pc to 3pc," *Addis Tribune*, May 31, 1996.

9. For a preliminary assessment of this new data on forest cover I am grateful to Peter Sutcliffe of Ethiopia's Woody Biomass Project.

10. For the fuller description of the expansion of plow agriculture, see McCann, *People of the Plow*, chap. 1–3.

11. See ibid., chap. 2.

12. See, for example, Richard Pankhurst, "The Foundation and Growth of Addis Ababa to 1935," *Ethiopia Observer* 6 (1962): 33–61; see also Ronald Horvath, "Addis Ababa's Eucalyptus Forest," *Journal of Ethiopian Studies*, 6 (1968): 13–19.

13. Horvath, "The Wandering Capitals of Ethiopia," *Journal of African History* 10 (1969): 205–19, cites Portuguese accounts as major sources for this argument. See Manoel de Almeida, *Some Records of Ethiopia, 1593–1646*, trans. and ed. C. W. Beckingham and G. W. B. Huntingford (London, 1954), 82.

14. W. Cornwallis Harris, *The Highlands of Ethiopia*, 2d. ed., 3 vols (London, 1844), 1:314–15; Charles Rochet d'Héricourt, *Voyage sur la côte orientale de la Mer Rouge dan le pays d'Adel et le royaume de Choa* (Paris, 1841), 115.

15. Rochet D'Héricourt, *Voyage*, 219; Harris, *Highlands of Ethiopia*, 2: 46–47, 150–51; and Charles Johnston, *Travels in Southern Abyssinia through the Country of the Adal to the Kingdom of Shoa during the Years 1842–43* (London, 1844), 2:63–64.

16. Douglas Graham, "Report on the Agricultural and Land Produce of Shoa." *Journal of the Asiatic Society of Bengal* 13 (1844): 254, 259.

17. See Christopher Clapham, *Transformation and Continuity in Revolutionary Ethiopia* (Cambridge, 1988), xi–xii. Clapham may have been the first to note the lack of empirical reference for the 40 percent figure. See also Almeida, *Some Records of Ethiopia*, 48.

18. John Martin Bernatz, *Scenes in Ethiopia*, 2 vols. (Munich and London, 1852), vol. 2, Plate VI (see illustration), indicated that Ankober households used banana trees for decoration and used their leaves for baking and storing bread. Banana trees only bore fruit in lower altitudes to the east below Ankober.

19. Guglielmo Massaja, *I miei trentacinque anni di missione nell'alta Etiopia*, 12 vols. (Milan, 1886), 11:39. Most of the town was destroyed, with only a single church left standing.

20. For slaves' role in wood gathering and wood as a royal privilege, see Johnston, *Travels*, 2:78–79; Harris, *Highlands*, 2:78; and C. W. Isenberg and J. L. Krapf, *The Journals of Rev. Mssrs. Isenberg and Krapf, Detailing Their Proceedings in the Kingdom of Shoa and Journeys in Other Parts of Abyssinia* (London, 1968), 120, 274.

21. Orazio Antinori, "Lettera del M. O. Antinori a S. E. il comm. Correnti Presidente dell Società," *Bollettino dell Società Geografica Italiana* 16 (1879): 403–4.

22. Graham, "Report," 263; Augustus Wylde, *Modern Abyssinia*, (Lon-

don, 1901), 399. Rosita Forbes, *From Red Sea to Blue Nile: Abyssinian Adventure* (New York, 1925), 138; Antonio Cecchi, *Da Zeila alle Frontiere del Caffa*, 2 vols. (Rome, 1886), 1:447–48; Helen Pankhurst, "The Value of Dung," in *Ethiopia: Problems of Sustainable Development: A Conference Report* (Trondheim, 1989), 75–88.

23. Graham, "Report," 289; Forbes *Red Sea*, 138. Jules Borelli, *Éthiopie méridionale: Journal de mon voyage aux pays Amhara, Oromo et Sidama, septembre 1885 à novembre 1888* (Paris, 1890), 98–101; Gustavo Bianchi, *Alla Terra dei Galla* (Milan, 1882), 158. Bianchi had to request forage for his pack animals directly from Menilek.

24. For an elaborate description of typologies and uses of dung in the highland farming system, see Helen Pankhurst, *Gender, Development, and Identity: An Ethiopian Study* (London, 1992), 90–6. Anderson and Gryseels report increasing use of dung in the 1980s around Dabra Zayt, the site of Ethiopia's most intense oxplow agriculture. See Guido Gryseels and Frank Anderson, *Research on Farm and Livestock Productivity in the Central Ethiopian Highlands: Initial Results, 1977–1980* (Addis Ababa, 1983), 10.

25. There is also important evidence that when the town and region supported a greater density of both population and authority, forms of intensified agriculture and natural resource management preserved rather than degraded the ambient natural resources. See McCann, *People of the Plow*, chap. 4.

26. Emilio Conforti, *Impressioni agrarie su alcuni itinerari dell'altopiano etiopico* (Florence, 1941), 157. For origins of coffee see Huffnagel, *Agriculture*, 204–5, and Raffaele Ciferri, "Primo rapporto sul caffé nell'Africa orientale italiana," *Agricolo Coloniale* 34 (1940): 135–44. While what many call "wild" coffee may be a naturalized domestic form characteristic of secondary forest growth, the large number of genotypes found in southwest forests argues for its origins there.

27. Daniel Gamachu, *Environment and Development in Ethiopia* (Geneva, 1988), 7, 12.

28. This curse is attributed to a nineteenth-century king of the Gera kingdom, deposed and blinded by an insurrection. Interview with Yalew Taffese, Gera, 29 November 1991. Field interviews for this research were conducted in Gera (Ethiopia) in November 1991 with support from the Social Science Research Council and the Norwegian NGO Redd Barna.

29. Cecchi, *Da Zeila*; Massaja, *I miei*; Enrico Cerruli, *Etiopia Occidentale (dallo Scioa alla frontiera del Sudan): Note del viaggio, 1927–1928* (Roma, 1933).

30. Cecchi, *Da Zeila*, 2:248. Cecchi arrived in December. The fields he describes as waterlogged were planted with maize in February, as is the practice there today.

31. Ibid., 2:256.

32. Ibid., 2:278.

33. Ibid., 2:277–78, 280–81, 279.

34. Ibid., 2:292.

35. Interview with Abba Dura Abba Bora, (Yukro) Gera, November 1992.

36. Massaja, *I miei*, 4:224.

37. Cecchi, *Da Zeila*, 2:382.

38. Interview with Abba Fogi Abba Chebso, Afallo (Gera), November 25, 1991.

39. West African virgin forest required 1,250 tons of vegetation removed for first cultivation; subsequent fallow clearance requires only 100 tons. See Chapter 6.

40. Cecchi, *Da Zeila*, 2:292–93; Paul Soleillet, *Voyages en Ethiopie (January 1882–October 1884): Notes, lettres et documents divers* (Rouen, 1886), 262–64.

41. Unlike in much of the north, slaves in Gera had an active role in field labor, cultivating for the royal houses as well as for landed aristocracy. For a comparative view of slaves in northern rural households, see James C. McCann, "Children of the House," in Suzanne Miers and Richard Roberts, eds., *The End of Slavery in Africa*. (Madison, 1988), 332–56.

42. Cecchi, *Da Zeila*, 2:277.

43. Names of plant species in Oromo and Amharic derive from a 1991 field visit and interviews with Shamsu Tewfik, Seyid Abba Karo (Yukro) and Abba Fogi Abba Chebsa (Afallo). For identification of Oromo names as Latin species see Wolde Michael Kelecha, *A Glossary of Ethiopian Plant Names* (Addis Ababa, 1987), passim.

44. Cerulli, *Etiopia Occidentale*, 156–57.

45. Interview of Abba Bor Abba Magal, Chala (Gera), January 21, 1990.

46. Cerulli, *Etiopia Occidentale*, 162.

47. Testimony to the mid–nineteenth-century lack of forest in Afallo parallels that for the other areas of Gera for which we have direct accounts. Interview with Abba Fogi Abba Chebsa, November 25, 1991.

48. For a thorough presentation of evidence on recent forest policies in Ethiopia, see Allan Hoben, "Paradigms and Politics," 1007–21.

49. I am grateful to Allan Hoben for sharing his research on this issue. See Hoben, "Paradigms and Politics," 17–20. It is also ironic that Swiss scholars such as Messerli and Hans Hurni have contributed very important geographic research on highland ecologies. See, for example, B. Messerli and K. Aerni, eds., *Simen Mountains, Ethiopia*, vol., 1, *Cartography and Its Application for Geographical and Ecological Problems* (Bern, 1978).

Chapter 6

Food in the Forest: Biodiversity, Food Systems, and Human Settlement in Ghana's Upper Guinea Forest, 1000–1990

Biodiversity defined most simply is the number and variety of species of life and the habitats in which they are found. Biologists and ecologists have tended to regard biodiversity as both the sum of individual species and as their sphere of interaction. Biodiversity is quantitative without being quantifiable. It is also a whole that we seem to value if it is rarer or larger in the number of its parts rather than simply larger or more homogeneous in its total number. Jane Guyer and Paul Richards point out that biodiversity consists of what we know and also an "ignorance zone" that includes that which is not yet known and probably never will be.[1] To this distinction we must add that biodiversity also has a place in time: it is a historical set of relationships between genetic materials—flora and fauna—that changes over time in ways that are chaotic and conjunctural with more infinite permutations of possibilities than human history. Indeed, one might say for the sake of argument that human history is a subset of the history of biodiversity. Biodiversity can also be a description of place, either a particular habitat (or "biome") or a specific geographic setting, like Africa.

The purpose of this chapter is to use biodiversity as a lens through which to view the history of a West African landscape, central Ghana, an area that comprises the frontier between the easternmost zone of the Upper Guinea forest and its adjacent savanna. This zone of forest is a biome at the center of a historical puzzle: How did state systems and complex human polities arise and develop in a protein-rich, carbohydrate-poor ecological setting that had, prior to the second millennium A.D. sustained only limited human settlement? What were the environmental foundations of political and economic power in the Asante empire that dominated the region from 1700 to 1900? Do forest ecology and biodiversity offer clues to the area's economic and social history? How did sources of food in the forest change over time, and what implications does this have for Ghana's future rural landscapes?

The history of central Ghana strongly reflects the engagement of human society with the vegetative cover of the Upper Guinea coast: the moist evergreen forest, the semideciduous fire zone on the savanna frontier, and the savanna itself. Key historical themes for this region include: settlement and state building in the forest (A.D. 1000–A.D. 1700), the invention of the forest fallow cultivation system, the Asante empire's biological landscape (1701–1823), cocoa's environmental revolution (1880–1920), forest reserves and maize fields (1900–96). In each of these overlapping environmental themes, the changing cover of vegetation and domestic plants has a central role in setting political contexts as well as overlaying the colors and textures of the historical landscapes. Mixed in with the confluences of human and landscape history is the central question of the history of Ghana's forest itself. A recent, and compelling, argument holds that Ghana's forests have not been disappearing at the alarming rate often cited by foresters and conservation advocates.[2] The new perspective holds that the actual historical size of the Upper Guinea forest has been seriously exaggerated and that human activity has, at times, actually preserved and expanded forest cover. This view seriously challenges the effects of fire, agriculture, and human labor on forest. It also potentially reverses the concept of "derived savanna" by arguing that local knowledge has historically regenerated a domesticated forest cover from savanna, that is, "derived forest." This chapter will feature that debate as an important interpretive template for the region's human history.

As a whole the crosscurrents of West African–Atlantic basin history infuse Ghana's history because of its resources in minerals (gold and bauxite), products of its forest (kola nuts, palm oil, and cocoa), and the aesthetic production of its peoples (*kente* cloth, brass gold weights, and the pageantry of Asante political and social ritual). The Asante empire that took root in the moist semideciduous forest in the seventeenth century was the largest and most successful polity in Ghana's past. By the time of its full flowering in the first decades of the nineteenth century, the Asante polity dominated several biomes, including the forest zones of central Ghana, the coastal savanna around Accra, and the savanna frontier that merged into the forest mosaic north and west of the Volta River. Asante and the Akan kingdoms that preceded it took part in the broad historical pageant that connected West Africa to the New World and the mercantile economy of the Atlantic basin beginning in the late fifteenth century. In addition to forming an international economy based on maritime technology, the formation of the Atlantic system involved the mingling of genetic resources (human gene pools, food crops, disease, plants, animals) and the massive redistribution of human labor in the form of slaves, migrants, and indentured servants who moved from Africa and Europe to the New World as well as moving within Africa.[3] West Africa's forests were a major actor within this play.

DISTINCTIVENESS OF THE UPPER GUINEA FOREST

Africa's rain forest belt straddles the continent at the equator north and south over several degrees of latitude with one exception: a sharp wedge of savanna that descends south and reaches the Gulf of Guinea between the Niger River (on the east) and the Volta River (on the west). Known as the Dahomey Gap, this area of savanna separates Africa's two great zones of forest. The Upper Guinea forest extends from Sierra Leone south and east to Togo, while the Congolia forest begins east of the Niger delta and connects the Cameroon and Congo (Zaire) forests. Africa's two major forest zones have many plants and animals in common, an indication of a historical corridor between them in moister epochs, but they also display a number of fundamental differences.

In general, the Upper Guinea forest is less species rich, has less rainfall, and has less species density than its Congolian sibling. Yet, it also has distinctive features that set it apart within the spectrum of rain forests globally. It has, for example, endemic varieties of tree frogs, squirrels, a pygmy hippopotamus, a range of specialized butterflies, and birds such as forest hornbills. The Upper Guinea forest's premier isolates, however, are its monkeys, six of which—including the rare Olive Colobus—have distinct forms west of the Dahomey Gap.[4]

While aspects of forest biodiversity such as frogs and squirrels hold great interest for conservation biology, other bits of evidence from

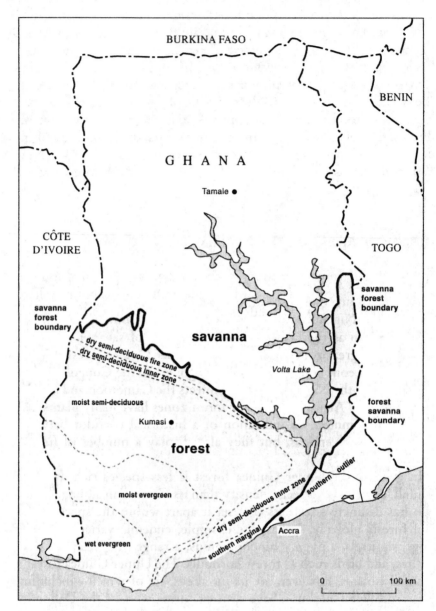

Figure 6.1 Map of Ghana.

evolution provide keys to environmental history. One of the primary signs of the long-term history of the Upper Guinea forest area is the wide variety of fauna that have versions both in the forest and the savanna but that are found only in very small numbers of purely forest forms. Even birds such as finches that subsist on grass seed have adapted their savanna habits to the forest habitat. This pattern of "derived" species in forest fauna—also found in squirrels, giant forest rats, and birds such as honeyguides—strongly suggests that dry epochs in this part of Africa have been more frequent than wet phases and that the dominant vegetative cover since the late Pleistocene has been savanna rather than forest. At the same time, the presence of specialized endemic frogs and gliding rodents (Anomalures, or flying squirrels) that cannot survive outside dense wet forest indicates that a small core of Upper Guinea's moist forest has managed to survive even the savanna's most advanced encroachments. Africa's history of radical climate change and sharply defined dry-wet seasons within each year makes the Upper Guinea forest unique among the world's rain forests (see Chapter 2).

Human adaptation to the Upper Guinea forest habitat comes late into the forest/savanna's evolutionary history (i.e., post-Pleistocene). Yet the forest's human population exhibited adaptive biological characteristics of its own. Recent work on the origins and distribution of the hemoglobin S, or sickle cell gene, indicate that it has its greatest presence in the blood of speakers of the Kwa language family (including Ghana's Akan peoples) who live in and around the Upper Guinea forest. F. B. Livingstone, a physical anthropologist, has simulated the diffusion of this gene into the African population and argued that it diffused into humans in central Ghana and the Upper Guinea forest fifty to sixty generations, or around 1300 years, ago.[5] This adaptation in humans that provided limited immunity to falciparum malaria clearly aided in the adaptation of human populations to a habitat shared with the *Anopheles gambia* mosquito, the primary vector for the disease. This evidence strongly suggests that ancestors of present-day central Ghanaians have occupied the forest zone since at least A.D. 700 to A.D. 1000.[6]

For the historical period covered by this book, however, evolutionary biology is less important than shorter-term patterns of climate, technology, the movements of human population, and the arrival of new plant varieties. In particular, the evolution of food systems and the forest environment in Ghana highlights the interaction of human production and forest ecology historically and places recent trends in food production into and within the context of the history of biodiversity.

A FOREST FALLOW REVOLUTION (A.D. 1000–A.D. 1700)

Forest ecology seemingly presents a set of insurmountable problems to the formation of centralized states. Indeed, the vast majority of Africa's historical empire states (e.g., Mali, Meroe, Great Zimbabwe, Aksum) have overwhelmingly centered themselves firmly on savanna ecologies that allowed easy communication and production of annual cereal grains that are storable, transportable, and easily divisible as tributary revenues. Obstacles to organizing the economy and political life of a large state in a moist forest setting would include problems of transport and communications; disease blocks to horses, cattle, and other livestock; and especially the need to provide adequate food supplies for concentrations of population.[7] While other historical conditions were also important, food supply was the *sine qua non* for concentrations of political power.

The Upper Guinea forest in central Ghana between the Pra and Afram Rivers was a relatively protein-rich setting for hunting and gathering. The forest offered various types of rodents—grasscutters, squirrels, forest rats—forest-adapted antelope, wild pigs, buffalo, and giant forest snails. T. C. McCaskie quotes A. W. Cardinal who in the 1920s described the importance of forest snails to the forest diet:

> Snails are one of the most important articles of food among the forest people. . . . The proper season for collecting the snails is at the beginning of the rains, and lasts about six weeks. Whole villages—men, women and children—migrate into the forest, leaving only the old and infirm to look after their homes. . . . It may seem extraordinary that one could get snails in such quantities. But once one has learned to detect them among the leaves, one will soon perceive thousands.[8]

Then as now, dried and smoked fish either from fresh water or carried from the coast may also have been a major protein source for forest diets.

The primary food dilemma for humans in the forest was the paucity of carbohydrates. Corms from wild yams and other tubers were part of the forest's biodiversity but existed too thinly on the ground to provide a caloric base for an army, bureaucracy, a population of town dwellers, or nucleated villages of taxpayers. While lowland rice was endemic to the Upper Guinea forest in Sierra Leone and Guinea, it seems not to have made an impact on the semideciduous forests of central Ghana. Other possible African grains such as sorghum and millet—crops that sustained the most dense savanna settlements— were long-season maturers that needed the sun and a long dry season to ripen. Yams were certainly a source of carbohydrates, a prestige food, and well adapted to forest soils, but were also a long-maturing, labor-intensive crop with a poor labor/maturity profile. In other words, West African indigenous yams were unlikely to sustain a growing forest population.

Forest soils offered good fertility and moisture on surface levels where leaf debris decomposed. Yet the forest's most precious commodity—sunlight—rarely reached the understory plants on the floor of a high canopy forest. The tallest trees used their command of the high canopy to win the competition for sunlight, allowing only shade-loving plants to occupy the vegetative understory in mature forests.

Solving this historical puzzle of the foundations of forest statecraft requires a fairly bold, even speculative, approach given the paucity of archaeological and historical evidence available. The distinguished historian of Asante Ivor Wilks has offered such a hypothesis rooted in both tantalizing shreds of historical evidence and the inexorable botanical facts of forest agronomy. His solution proved to be a description of the Akan people's historical evolution of forest fallow agriculture linked to the conjuncture of the emerging Atlantic economy.

Wilks's hypothesis tackles the forest dilemma with two related sets of evidence and arguments. First, he established the historical parameters of the forest-clearing process by extrapolating from recent evidence on farm size, the caloric needs of an Akan family unit, and the requirements of labor for clearing forested lands. In the former case he hypothesized that the needs of cultivating forest plots and the average caloric needs of a family of five suggest 6 hectares (15 acres) as the minimum farm size to support an Asante household. Assuming a fifteen-year fallow rotation such has existed in more recent times in central Ghana, each 1-hectare plot would be in production three years out of every eighteen. Farm sizes on this order would thus have been capable of maintaining soil fertility and sustaining a population in central Ghana of around 500,000 or 130 persons per square mile.[9] This type of farm-producing grains, tubers, and vegetables would in the aggregate have been able, in theory anyway, to support the density of population generally attributed to the Asante empire at its highest point of political, economic, and military influence in the early nineteenth century.

The second part of Wilks's hypothesis addressed the forest/farm ecology of how such farms emerged from the region's moist semideciduous forest. Fundamentally, the issues are the twin dilemmas of (1) how to remove high canopy, primary forest vegetation to allow sunlight to penetrate to food crops at ground level and (2) the need to control the menace of vegetative regrowth choking farm fields after clearance. The figures are staggering: clearance of a single hectare of virgin forest called for removing 1,250 tons of moist vegetation by using the cutlass, bill hook, and fire as the primary tools. It was hot, dangerous, and arduous work to remove what J. Dupuis, an observer of Asante forest clearance methods in the early nineteenth century, called the "cumbersome growth of fibrous stems and vines, mixed with other plants of a watery nature."[10]

The critical feature of Wilks's equation, however, was not so much the daunting weight of biomass to be hacked, cut, uprooted, dragged away, and burned, but the fact that after the first clearance of 1250 tons, subsequent clearance of the same plots after a fifteen-year fallow was only 100 tons! The key task historically then was breaking the forest canopy's monopoly on sunlight through the first-stage removal of the canopy and the choking understory of trees, bushes, and vines. Once cleared of the forest's primary canopy, an agricultural economy could emerge to expand the agricultural frontier into the forest; later stages released labor for crop cultivation, military service, construction, and so forth.

Nineteenth-century observers as a whole tell us that an ecological revolution had already taken place by the time of their arrival. By the time of Dupuis's 1820 visit to the Asante heartland, the forest fallow system was well entrenched and much of the land clear felled, leaving a landscape akin to "the country gardens of Europe."[11]

How was this cleared and cultivated landscape achieved given the initial costs of clearance in labor? Wilks answers the labor question with an intriguing set of oral and documentary historical evidence that points to the fifteenth and sixteenth centuries as the time of a crucial conjuncture that drew new human populations into the central forest

Figure 6.2 Felling a mahogany tree, central Ghana.
(Courtesy of Dioff, African Studies Program, University of Wisconsin-Madison.)

and provided the social mechanisms (matriclans or *abusua*) to integrate them.

The first European contacts with the West African coast in the fifteenth century in search of gold found that there was already an active slave trade that brought captive labor from the Niger delta to the Ghanaian coast. As the Portuguese traders quickly learned, the easiest way to obtain gold from the mines of central Ghana was to transport slaves from elsewhere on the West African coast to the new (founded in 1482) Atlantic entrepôt at El Mina. This influx of labor to the central Akan region from the Portuguese coastal trade, in Wilks's

Figure 6.3 El Mina Fort and surf port, est. 1482. Photo by author.

view, was only a smaller version of large-scale labor imports from the north through the trade networks with the Mali empire that had continued its own need for gold to foster its trans-Saharan trade interests.

The new population influx into the forest presented problems of how to integrate new persons into its labor system without creating an unwieldy servile class. What seems to have evolved historically was an open social system based on matrilineal clans (*abusua*) and a social philosophy emphasizing assimilation. Akan oral traditions situate the foundation of their distinctive matrilineal clans not in a metaphor of the creation of humankind but as a specific set of events in the forest several generations prior to well-known rulers of the seventeenth and eighteenth centuries. It was at that historical moment that Akan oral traditions describe a group of ancestresses who descended from the sky on chains and founded the great matriclans of the Akan people in the forest.[12] In other words, as Wilks pointed out, the Akan people were in the process of expanding their population and labor capacity at about the same point that they developed a social system that encouraged population growth. Gold extracted from forest mines provided the means to attract labor and link forest ecology to the expanding Atlantic world.

Wilks's critics have pointed to evidence of forest settlements from as early as A.D. 1000 and signs that populations in the region had built defensive earthworks (presumably against raids) and then disappeared, implying depopulation and "genocide" rather than net growth in labor. The evidence of the sickle cell trait mentioned above has also suggested a forest settlement period earlier than the fifteenth and sixteenth centuries.[13] None of this evidence, however, is inconsistent with a general view that existing forest populations began from a low population density foraging economy that grew rapidly and underwent a social transformation in the half millennium before the birth of the Atlantic economy. In fact, the critiques of Wilks's thesis would seem to indicate that management of the forest ecology was already developing before Akan expansion; new labor was a means to enhance a building repertoire of local knowledge of a forest ecology, of protein collection, and production of carbohydrates.

Arguments for extensive pre-1500 forest settlement nevertheless, ignore an ingredient that points directly to a take-off point of around 1500: the arrival from another world tropical forest biome—the forest zone of Central and South America—of new food crops ideally suited to provisioning an expanding set of forest polities. New World food crops—cassava, cocoyam, cowpeas, and above all, maize—brought by Portuguese ships to provision their coastal fortresses spread quickly beyond the fortress gardens and created and sparked an agricultural carbohydrate revolution that allowed peoples of the Upper Guinea forest to feed a dense, growing population and foster the elaboration of an elite political class, royal courts, and a standing army. The confirmation of Wilks's hypothesis on the fifteenth-century "big bang" in the forest of central Ghana is the fundamental role of cassava and maize as the frontal edge of a human assault on the forest landscape using "forest fallow," an accumulated repertoire of knowledge and revolutionary new plant germ plasms, to convert the forest's biomass and energy into carbohydrate calories.

Observers of the dynamics of the forest fallow farming system have described elements of it since the early nineteenth century, but its fullest elaboration appears in the insightful field work of Kojo Amanor, an anthropologist and geographer. Working in the 1990s, Amanor, however, describes the system as a recent adaptation rather than historical knowledge despite evidence of its practice in the earliest European accounts of the Asante empire. Fairhead and Leach (see Chapter 4) argue from their part of the Upper Guinea zone that forest

Figure 6.4   Forest fallow: maize stalks in foreground and forest succession at rear, eastern province, Ghana, 1975. Photo by author.

fallow systems change landscapes by regenerating forests. In their view what Amanor describes as recent forest farm responses to degradation of the eastern forest transition zone may well fully detail the system in place throughout central Ghana after sixteenth century when forest farmers and gatherers added maize and cassava to specialized niches within a forest cultivation system that had evolved much earlier.[14]

Ghana's Upper Guinea forest cultivation system derives its rhythms and limitations from two seasons of rainfall and a season of dry winds (the harmattan) that blows dry air and dust from the Sahel toward the Gulf of Guinea. The heavy rains take place between March and early July; a second, smaller rainy season begins in September and lasts until October and allows the cultivation of a second plot. The farm cycle begins with clearing fields for the primary farm between December and January. Farmers use cutlasses to slash and remove the understory of shrubs and ground cover. Farmers then cut major branches from large trees (pollarding) to open the canopy cover and add leaf debris as a mulch to the soil's surface. With the beginning of rains in March, farmers place leaf debris, smaller branches, and other vegetation, now dried, into piles for a controlled burning. The use of fire reduces the weight of the forest biomass to an ash and destroys insects and small weeds. Burning also releases phosphorous and singes the leaves and branches of the largest trees, debris that may fall to the ground, adding organic matter to soil, enhancing the layer of mulch, and increasing sunlight's penetration to the forest floor. This process of clearing, burning, and pollarding, in effect, converts the energy stored above ground in forest vegetation into soil nutrients to feed food crops.

With the first rains of March, farmers plant yams in mounds near small trees that serve as stakes for the plant's emerging tendrils. At that point, fire has already created leaf fall that allowed sunlight to reach the yam mounds. With the March rains fully established, farmers plant maize by using minimum tillage techniques to keep mulch and soil moisture in place. Maize thus received the moisture essential for its early growth and tasseling. Two weeks later, farmers plant cassava sticks between the germinated maize; cocoyam corms preserved in the soil begin to sprout. As these crops grow, a dense vegetation emerges, with maize leaves leading the way toward the intense sunlight they require. The leaves of the young cassava and cocoyams closer to the forest floor cool the maize roots and protect the forest soils from rainfall impact and direct sun.

After the October maize harvest, cassava and cocoyams remain in the field, offering shade to suppress weeds and protect the soil. Both crops require at least a year's maturity, but cassava can remain in the ground for two to three years, providing food as the fallow cycle begins. Plantain suckers, a new arrival from Asia by the sixteenth century, may also have been added to this mix of intercropped plants on forest clearings. On a single cleared plot, farmers could plant maize in three successive seasons and allow the cassava plants to continue to grow. In Amanor's field study, farmers also opened a second farm plot at the end of August to take advantage of the second rainy season. This farm was often only for a single planting of maize that could come to harvest within the short rainfall period and provide a source of food prior to the maturity of the tuber crops.[15] Maize's capacity to provide a farm's second harvest in a single season offered a strategic boost to local food supply.

As Wilks's thesis indicates, the management of fallow land was a crucial aspect of the system's resilience. After the maize harvest of the main farm, farmers began managing the succession of regrowth, a process that drew on accumulated knowledge of biodiversity and soil behaviors built up over generations of forest occupation. The labor savings of managing fallow as opposed to opening up new or secondary forest are clear from Wilks's calculations cited above. Farmers on fallow plots deliberately preserved small, pioneer forest species that they knew had nutrient recycling properties. Amanor cites eight species of trees and nine shrubs and herbs that local farmers recognized as having soil enhancing effects.[16] Fallow management was thus not abandonment of old plots but a careful and conscious selection process that preserved particular species and removed others to allow a sustained production of food. This food supply, in turn, released labor to push the frontiers of forest settlement and support development of the art of politics and statecraft.

The forest fallow system that appeared in the written historical record only in the early nineteenth century (Portuguese traders had never seen it firsthand) was, therefore, the end result of generations of experimentation and observation, first through the efforts of hunting and gathering forest dwellers in the early second millennium A.D. and then as part of an agricultural transformation that drew new labor into the forest to break the logjam of primary clearance. The engine that pushed the "forest fallow" revolution was the arrival of three New World domestic plants that occupied strategic niches in the forest fallow system. Maize provided a new cultigen that offered something new—an early maturing food source that provided carbohydrates by the end of the rains and that required much less labor than yams did. Maize also offered a second harvest, while cassava complemented maize's early yield and double crops by being a low-labor crop with the capacity to remain stored in the ground for extended periods. Moreover, cassava could tolerate poor soils, drought, and neglect; cassava leaves also provided understory shade under the sun-loving maize leaves.

The expansion of forest fallow cultivation into central Ghana's forest therefore brought about an early change in the forest's biodiversity. Even as population in central Ghana declined through the nineteenth century (the result of a dispersal to new frontier zones) following the slow dissolution of Asante hegemony and the rise of a colonial *pax Brittanica*, the secondary regeneration of forests bore the marks of forest fallow's selection process. In the Kakum National Park forest (near Cape Coast) at the end of the twentieth century, there is still evidence of human occupation and forest fallow cultivation in the form of enriched soils and the residue of plant succession and selection recognizable to park rangers and hunters.

THE BIOLOGY OF ASANTE STATE POWER (1700–1823)

The evolution of a forest food supply and forest fallow agronomy was one of several factors, including Atlantic trade expansion, dynamic leadership, and access to firearms, that contributed to the rise of the Asante empire state in the eighteenth and early nineteenth centuries. When Osei Tutu Kwame (later nicknamed Osei Bonsu) ascended to the stool as Asantehene in 1804, the empire was at the height of its economic and military power. It stood at the center of an elaborate network of provinces, tributary states, and European clients on the coast that provided revenues, ritual obeisance, and labor. Social and political life revolved around the Asantehene's court in Kumasi and in the great matriclans that had provided the fictive kinship ties and social mechanisms to absorb rapidly new populations in the period of expansion into the forest environment. By the early nineteenth century, Asante hegemony reached into the savanna to the north and to the Gulf of Guinea, thus encompassing several biomes that complemented the forest that had served as its creche.

The 1804–23 reign of Osei Bonsu also offered a detailed set of historical snapshots of the central Ghanaian landscape. As part of his essay into the labor and land history of Asante, Wilks has surveyed the observations of European diplomats and travelers who described vegetation and land use.[17] From these accounts we can discern aspects of change in the region's biodiversity. At its center, the forest fallow system had already transformed its piece of the Upper Guinea forest around the royal capital at Kumasi into an ordered, intensively cultivated human landscape. Kumasi by then was a town of around 20,000 souls that swelled with crowds of visitors and dignitaries on days of ritual and political significance, such as during the *Odwira* "first fruits" celebration in October. Thomas Bowdich who visited the capital in 1817 described the plantations that surrounded the town for several miles:

> There were continued plantations of corn [certainly maize], yams, groundnuts, terraboys, and encruma: the yams and groundnuts were

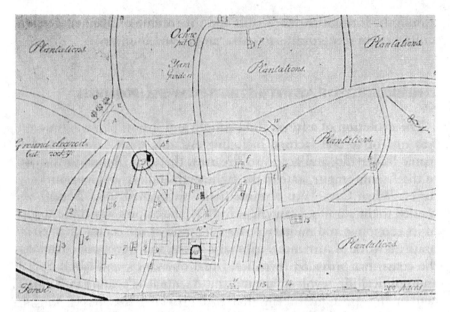

Figure 6.5 T. E. Bowdich's map of Kumasi land use, 1817. From T. E. Bowdich, *Mission from Cape Coast to Ashantee* (London, 1819).

planted with much regularity in triangular beds, with small drains around each and carefully cleared from weeds. [The yam plantings] had much the appearance of a hop garden well fenced in, and regularly planted in lines, with a broad walk around, and a hut at each wicker gate, where a slave and his family resided to protect the plantations.[18]

This description makes it clear that in the intensely cultivated zone around Kumasi the forest fallow system had evolved into a form of labor-intensive (using resident farm labor) permanent cultivation with rotations of maize, groundnuts, and perhaps composted crop residues to balance nitrogen and maintain soil fertility. Descriptions of the farming system do not mention major sources of animal protein, so we may assume that hunting and forest collections provided meat from rodents, antelope, and snails drawn from adjacent forest areas.

Wilks speculates that the Asante landscape included three "levels" of land use within the forest zone and a further area of what he calls "derived savanna."[19] Each of these levels represents a version of forest fallow cultivation. In some areas the state had decreed sacred groves preserving forest species that provided medicine, building materials, or settings of ritual importance. Perhaps it was one of these preserved sacred groves that Dupuis saw near the capital in 1820 and described as having "thickets and entanglements," and "characteristic gloom."[20] In general, however, there was a landscape pattern that displayed intensive cultivation of food crops around key political centers (which Wilks calls Level I); short-fallow rotation cultivation and secondary forest growth (Level II); and frontier areas of new clearing with forest fallow cultivation in canopy forests (Level III). In the latter case Bowdich stated that wealthy Asante at Kumasi continued to send dependent workers to settle the forest frontiers and open new farms by using what was by then a well-established set of tools and crops to break through the semideciduous forest's dense primary canopy and establish the forest fallow system.[21] In each of these zones the human impact on biodiversity approximated the earlier pattern: clearing, planting of seeds and tubers, selection of favored fallow tree species, and improvement of soils around settlement sites. The moist evergreen forest zone in the southwest that had sustained a number of endemic species of flora and fauna may have been spared the major impact of human settlement because of its distance from political centers, along with areas within the central forest that presented some type of resistance—areas of steep slope, swamp, or ritual significance—to human settlement.

Forest fallow itself implied a deep accumulated knowledge about plants, animals, and forest chemistry that offered sources of food, poison, building materials, and the tools of soil management. The Upper Guinea forests' characteristic (and relative) homogeneity of species allowed local knowledge a wider field of play across all of the zone. Asante expansion in the mid–eighteenth century had brought savanna areas under its control. This contact probably increased the importation of domestic livestock as sources of protein, especially Ndama cattle that could survive entry into the tsetse zone. More important to food supply and Asante tastes was probably the regularization of movement of maritime protein such as dried, smoked fish and crustaceans into forest markets along the major roads to Cape Coast, Accra, and other coastal entrepôts.

The success of New World cultigens in the forest fallow revolution had established a set of staples within forest diets. If yams bore prestige and commanded considerable attention in royal plantations, maize, plantain, and cassava probably provided the bulk of the calories. As a grain rather than a tuber, maize had the advantage to the state of being storable, divisible, and transportable.[22] Farmers also ap-

Figure 6.6   Maize cultivation in northern Ghana, 1975. Photo by author.

parently favored maize for its adaptive capacities. Through observation and seed selection, farmers by the early twentieth century had produced maize cultivars with characteristics specific to various biomes. Local maize varieties preferred by farmers were floury semident with closed tips and with maturity dates ranging from 95 to 120 days. These characteristics made local varieties more easily handmilled (semident is starchy and fleshier), resistant to postharvest loss (closed tip discouraged insect damage), and synchronized with local rainfall patterns and rotational practice.[23]

We know little about the actual date of introduction of maize to the Upper Guinea forest, but one report states that the Asante introduced it to the savanna in the early nineteenth century. At that point it had existed within the forest fallow system for several centuries at least. By 1820 Asante agronomy and statecraft had intertwined to the extent that the consumption of maize became the special responsibility of the Asante commissariat that provisioned the Asantehene's army in the field. As the Asante kingdom projected its military and political authority north onto the savanna, they also introduced maize to its drier northern hinterlands where farmers selected seed stock to develop an early maturing (ninety-five days) cultivar that served as

Figure 6.7 Cassava for sale, Accra market, 1975. Photo by author.

both human food and livestock forage. Wilks argues that a decline in the popularity of maize in Asante agriculture between the mid–nineteenth and mid–twentieth centuries was a direct result of the decline of Asante military fortunes.[24] Wilks also argues from historical evidence that maize was far more popular in central Ghana in the early nineteenth century than in the mid–twentieth century because a new cultivar of cocoyam from the West Indies (*Xanthosoma mafaffa*) arrived in the 1840s and replaced both maize and the less digestible older yam variety *Colocasia esculenta*. By the post–World War II era, imported rice and wheat had also increasingly penetrated urban markets.

By the mid–nineteenth century, central Ghana's forest landscapes were more than ever necessarily human landscapes. Doubtless, the patterns of forest penetration and settlement in Asante were strongly affected by population densities. Wilks and others have argued that the population of greater Asante may have reached 500,000 in the first decade of the twentieth century and then declined through the rest of the century. Population dispersals or the end of the slave trade may have reduced the concentrations of labor necessary to sustain a tight forest fallow and resulted in an expansion of secondary growth in nonurban zones. Those forests, however, continued to bear the marks

Figure 6.8 Cocoa cultivation, central province, Ghana, 1975. Photo by author.

of human activity in the form of plant selection, use of fire, and introduction of new species.

COCOA REVOLUTION (1880–1920)

Anthropologist Kojo Amanor describes the mid- to late-nineteenth century in central Ghana as the beginning of a pioneer frontier settlement process in which farmers moved into new unsettled portions of the forest to use their natural resources as extractive windfalls. The first stages of this expansion were the forest fallow expansion described above. By the mid–nineteenth century, global markets for industrial commodities stimulated the collection and cultivation of nuts of the oil palm, a plant that grew wild but also offered quick results from plantation. Collecting palm nuts from the forest with family labor supplemented farm income and linked forest ecology to the world economy. By the 1880s, however, farmer response from West Africa had overwhelmed the palm oil market.[25] With the arrival of new petroleum lubricants and new Asian producers, demand collapsed and a new opportunity emerged that encouraged cocoa as a new frontier movement into the forest ecology.

Local Ghanaian legend recounts that in the mid–nineteenth century a Gold Coast sailor named Tettèh Quarshie brought the first cocoa seedling, another New World plant, from the island of Fernando Po. By 1880 cocoa had found a world market demand and a place within the economy of forest frontier settlement in Gold Coast economy beginning in the east but moving rapidly westward toward the moister forest, often complementing oil palm cultivation. Polly Hill's classic study of the economics and geography of cocoa production in Ghana suggests that Ghanaian farmers began the systematic cultivation of cocoa trees in "virgin forest" in the eastern region between Accra and Kumasi in the early 1890s.[26] By 1894 there was migration to forest frontiers to establish cocoa farms within the forest canopy

Unlike the oil palm that produced a commodity for both international markets and local use (as edible oil, palm wine, and distilled spirits), cocoa was a pure commodity that linked small farms in the forest to a cash economy and a world market.[27] Cocoa production that had been around 500 metric tons in 1900 increased to 22,631 tons in 1910 and to 200,000 tons by 1930.[28]

Vibrant markets, the *pax Brittanica's* open frontiers, and a ready population of migrant farmers laid the foundation for a new movement of human imprint on forest ecology. The ecological dynamics of the cocoa frontier allowed a rapid expansion of a new form of clearance and selection in stages. Cocoa's spread into the forest added the agronomy of a fruit tree to the food crop forest fallow system. Amanor has pointed out that the addition of the tree crop to the fallow system mimicked the forest through creation of a shade canopy to control the understory of weeds and protect the soil humus from direct sunlight. Where farmers intended cocoa seedlings to replace primary forest, clearing was identical to the classic forest fallow sequence: farmers followed clearing and burning with three years of maize, yams, and vegetables. Next, the shade crops cassava, cocoyam, and plantain provided cover for young seedlings and a sequence of food for the five-year period before cocoa seedlings produced a marketable crop. When mature cocoa trees covered the forest farm, the process was complete except for seasonal slashing of weeds and harvest. The farm enterprise could then move into another forest plot and repeat the process, moving further into the moister forest frontier to the west and drawing migrant labor with it. Migrants who developed cocoa farms on landlords' plots retained the food crops they produced, while landlords reaped the longer-term benefits of the cocoa fruits and ownership of a mature cocoa farm.

Figure 6.9   Road through the forest, Ghana, 1975. (Courtesy of African Studies Program, University of Wisconsin.)

The frontier economy of cocoa cultivation moved progressively into new zones ecologically capable of supporting it, necessarily and simultaneously creating a new human landscape and a new forest ecology. The arrival of railway stations at Nsawan and Pakro in 1910 and 1911 allowed the Gold Coast's forest and farmers to attain the status thereafter as the world's leading cocoa producers. After 1918 the expansion of motorized lorry transport allowed world markets to penetrate fully the forest biome.[29] With transport costs falling and post–World War I world market demand rising, forest land attained a new economic value. Production on forest lands by new migrant populations outstripped its value as a source of extraction for local peoples.

Over time, farmers were thus able to add new dimensions to managing forest ecology. A striking example is the evolution of managing ant populations.[30] The Upper Guinea forest, like rain forests elsewhere in the world, is home to a kaleidoscope of insects. As many as 300 species of ants may occupy a single cocoa plantation, often dominated by species that damage cocoa trees. Over time, farmers have learned to manage ant ecology by studying species behaviors and manipulating nests to drive out or suppress damaging types.[31] In other cases, farmers have learned to balance the shade tolerance of cocoa and food crops to integrate the cocoa canopy with the needs of plant under-stories. Farmers interviewed by Amanor in the Krobo district dis-played a sophisticated knowledge of trees and shrubs that balanced soil chemistry and allowed regeneration of fertility in forest plots in and around human cultivation.

In the early 1940s the ecology and economy of the forest changed again. Mature cocoa farms in the eastern region began to show signs of crisis in the form of sudden deciduousness of normally evergreen cocoa trees and the spread of the devastating swollen shoot disease. In addition to the trees that died directly from the disease from the late 1930s on, by 1961 the Ministry of Agriculture has destroyed 105 million trees.[32] Contemporary accounts and farmers interviewed by Kojo Amanor also reported increasing dryness of the climate in the eastern range of cocoa cultivation. With the loss of cocoa trees from disease and fires, eastern farms in the dry semideciduous forest zone reverted to forest fallow food cultivation of maize and cassava mixed with a minor revival of oil palm cultivation.[33] The overall decline in rainfall and spread of disease on the old mature cocoa landscapes pushed the forest frontier into new areas and foreshadowed a new form of human land use on forest ecology. Beginning in the mid–nineteenth century, the cultivation of oil palm and then cocoa had existed in a kind of symbiosis with the production of food crops within the forest fallow system, a relationship that maintained shade cover of forest soils. However, the conversion of old cocoa farms to food production in the post–World War II period and a political movement to place forests within protective reserves ushered in a new period of landscape history in central Ghana's piece of the Upper Guinea forest.

FOREST RESERVES AND MAIZE FIELDS (1908–1996)

Early in the twentieth century, colonial observers of the environ-ment expressed concern over what they assumed to be the accelerating loss of primary forest (see below). In 1908 the British colonial gov-ernment of the Gold Coast surveyed the colony's forest resources and established a Forestry Department to control its exploitation. Then in 1927 the colonial government formally set aside certain lands as forest reserves. By 1939, 1.5 million hectares of land—almost 20 percent of land within the moist forest zone—fell under government protec-

Figure 6.10 Timber transport, Birimi Logging and Lumber Company, Kumasi, 1997. Photo by author.

tion.[34] The forest reserve system had two fundamental implications for central Ghana's ecological history: (1) it brought important natural resources under the direct control of the state, and (2) assumptions about forest cover built into the plan contributed to the later exaggerated estimates of forest losses in the post–World War II period.

With the assumption that all reserved areas were under primary forest, foresters and policy makers have calculated alarming figures for deforestation over the course of the twentieth century. Such estimates of loss state that Ghana's closed forests once occupied over 8 million hectares but have declined to 1.9 million by 1980. Using similar figures, the World Bank in 1988 estimated that Ghana's closed forest had declined at an annual rate of 75,000 hectares since 1900.[35]

More than a reality of deforestation, the colonial impulse to reserve forested land reflected a "misreading" of the historical ecology and a desire to protect key resources for state management (a policy later adapted by many African governments at independence). A recent close examination of the origins and validity of data on forest cover in Ghana effectively challenges such estimates. James Fairhead and Melissa Leach's most recent work on forest history in West Africa

Figure 6.11 Timber milling, Kumasi, 1997. Photo by author.

points out several misleading historical assumptions about Ghana's forest resources:

1. Many of the lands classified by colonial officials as "closed forest" were, in fact, clearly secondary growth, former oil palm farms, or canopy-covered cocoa estates.

2. Population declined or dispersed in the mid–nineteenth century, leaving many former cultivated lands and settlements to regenerate into secondary forests.

3. Many estimates of forest cover assumed that all land in forest reserve areas consisted in fact, of closed high-canopy forests. In fact, boundaries for reserves often reflect watersheds, fire zones, or areas of former cultivation.[36]

Beyond the miscalculation of historical forest cover on environmental resource policy in Ghana, this research sheds new light on historical landscapes themselves. Ghana's forests in the nineteenth and twentieth centuries were, in fact, a patchwork of human settlement, cultivated fields, sacred groves and regenerating secondary growth.

FROM COCOA FOREST TO MAIZE FALLOW

In addition to the evolution of forest management and cocoa cultivation over the course of the twentieth century, Ghana's agricultural economy has shown increasing signs of a more profound change in its forest landscapes. Declining cocoa production in some areas, urbanization, and increasing food producer prices have created new incentives for smallholder farms to rediscover the cultivation of maize, this time as a commercial commodity. As stated above, maize had been an important food crop for the Asante state in the early nineteenth century. With the adoption of new varieties of cocoyam and the expansion of mature cocoa farms, maize's popularity declined by the early twentieth century.

In the final quarter of the twentieth century, however, maize has again soared in its popularity with farmers and consumers. Ghanaians consume maize as *kenkey* (fermented maize meal boiled in maize husks or plantain leaves), as *tuo zafi* (a maize porridge), or as ears roasted on charcoal. Between 1975 and 1991, for example, the area planted in maize increased from 319,700 hectares to 610,400 hectares, an increase of 125 percent. More dramatic, between 1979 and 1995 total maize production increased 230 percent (i.e., yield has risen from 1 to 1.5 tons per hectare, an increase of 50 percent).[37] Moreover, two-thirds of Ghana's maize farmers sow maize as a cash crop and not merely as a household food supply. By the turn of the twenty-first century, maize will have consolidated its status as Ghana's primary food source.[38]

Maize has a broad appeal to farmers because of its agronomic characteristics: it requires less labor for a quicker yield than any other food source. It is ideally suited to northern savanna zones at the edge of the forest, readily replaces cocoa trees damaged by fire or disease, and appeals to stressed urbanites who seek a low labor food crop to grow between houses and beside city streets. Improved varieties released by commercial seed producers have continued to emphasize high yields and early maturity.

In the 1990s a new program, the Sasakawa/Global 2000 project, has

Figure 6.12   New timber regulation, Ghana, 1997.

introduced maize as the linchpin of a green revolution package for Ghana's farmers. The package includes improved or hybrid seed, fertilizer, row planting, herbicide, and no-tillage techniques.[39] New varieties of maize being distributed include Obatanpa, a QPM (Quality Protein Maize), that provides lysine, a critical amino acid missing from maize, but necessary to releasing complete proteins. This program even includes special provisions for intercropping of maize and legumes (beans and peas). Station trials of this package have yields per hectare that are 100 to 150 percent higher than current on-farm averages.[40]

The rapid expansion of annual cropping of maize and legumes in Ghana has specific implications for the forest biome and its biodiversity. First, the increasing cultivation of maize by using green revolution techniques implies an industrialization of agriculture and food supply, involving commercial seed producers, the need for farmers to obtain seed from the market instead of their own fields. This also means that local varieties, and their diverse germ plasms, will begin to disappear, requiring the establishment and funding of gene banks to collect and store local seed types for use in breeding new improved species with resistance to disease, drought, and pests. Moreover, fields will need to be under permanent cultivation, which means the use of fertilizers to replace nitrogen and herbicide to suppress the aggressive weeds, especially striga, that invade fields after its initial clearing.

Second, maize is a notorious lover of direct sunlight and therefore demands the permanent removal of forest cover, both primary and secondary. Permanent plots of high-yielding, protein-rich maize will need an exclusive canopy of sunlight: forest canopy and high-yielding maize cannot coexist. Ironically, the cost of fertilizers under this new tyranny of maize will do much to determine the forests' future. With low, that is, subsidized, fertilizer prices, farmers can maintain permanently cultivated plots by using intercropping, row planting, and no-tillage strategies. By contrast, higher prices for fertilizers (i.e., the loss of farm subsidies under World Bank structural adjustment), however, encourage farmers to clear new lands, that is, forest land, to provide the natural fertility to sustain nitrogen-hungry maize. The growth of Ghana's cities, towns, and wage labor population will undoubtedly increase in the twenty-first century. The spread of the sustainable maize farm will continue, with dire consequences for both the forest fallow cultivation system and the forest canopy.

Figure 6.13   Urban maize cultivation, Accra, 1997. Photo by author.

CONCLUSION

The Upper Guinea forest of central and southern Ghana reflects a long and complex history of interaction between forest and savanna biomes as well as the impact of human settlement over the past millennium. In the first instance, the peculiar quality of the Upper Guinea forest emerged from alternating dry and wet epochs and annual seasonality that shifted the boundaries of forest and savanna landscapes and fostered flora and fauna's adaptation to two volatile biomes. The pattern of biodiversity that evolved, therefore, displays less endemism and more widely distributed species.

Within the last 1000 years, human occupation of the forest/savanna frontier has brought about spurts of rapid change in the composition of plant species, the admixture of New World food and commercial plants, and, in the most recent phase, landscapes of annual crops (especially maize) that have replaced forest fallow. The transformation of Ghana's landscapes and biodiversity has moved almost in lock step with the evolution of its human food supply. Seasonal rhythms of harvest, planting, and clearing have replaced the evergreen forests in the most recent phase of change that has included human struggles to secure sustenance from forest ecology. In the future, the need for forest biodiversity conservation will stem not from human action of the past but from the imperatives of the world economy of development in the present.

NOTES

1. Jane Guyer and Paul Richards, "The Invention of Biodiversity: Social Perspectives on the Management of Biological Variety in Africa," *Africa* 66, (no.) 1, (1996): 2.

2. James Fairhead and Melissa Leach, *Reframing Deforestation: Global Analyses and Local Realities—Studies in West Africa* (London, 1998). My thanks to the authors for sharing a draft copy of their manuscript.

3. For a fuller description of the "Columbian Exchange," see Alfred Crosby, *The Columbian Exchange: Biological and Cultural Consequences of 1492* (Westport, 1972).

4. Kingdon, *Island Africa*, 105–7.

5. F. B. Livingstone, "Anthropological Implications of the Sickle-Cell Gene Distribution in West Africa," *American Anthropologist* 60, no. 3 (1958): 533–62, and F. B. Livingstone, "Who Gave Whom Hemoglobin S: The Use of Relative Restriction Site Haplotype Variation for the Interpretation of the Evolution of the B^s-Globin Gene," *American Journal of Human Biology* 1, no. 3 (1989): 289–302. Both articles are cited in Norman A. Klein, "Toward a New Understanding of Akan Origins," *Africa* 66, no. 2 (1996): 249–51.

6. For archaeological evidence of early settlement, see Merrick Posnansky, "Prelude to Akan Civilization," in *The Golden Stool*, ed. Enid Schildkrout, (New York, 1987).

7. For the relationship of state building and cereal crops, see McCann, *People of the Plow*, 153–54. Buganda, where the plantain served as the major food source, may be the exception to this pattern.

8. A. W. Cardinal, *In Ashanti and Beyond* (London and Philadelphia, 1927), 78–80, quoted in T. C. McCaskie, *State and Society in Pre-Colonial Asante* (Cambridge, 1995), 29.

9. Ivor Wilks "Land, Labour, Capital and the Forest Kingdom of Asante: A Model of Early Change," in ed. Jonathan Friedman and M. J. Rowlands *The Evolution of Social Systems*, (Pittsburgh, 1978), 497–501. This research appeared in a somewhat revised version in Ivor Wilks, *Forests of Gold: Essays on the Akan and the Kingdom of Asante* (Athens, 1993), 41–90. For population estimates, see Ivor Wilks, *Asante in the Nineteenth Century: The Struc-*

ture and Evolution of a Political Order (Cambridge, 1975), 87–93. See also M. Johnson, "The Populations of Asante, 1817–1921: A Reconsideration," and Ivor Wilks, "The Population of Asante, 1817–1921: A Rejoinder," *Asantesem: The Asante Collective Biography Project Bulletin* 8 (1978): 22–28 and 28–35.

10. Wilks, *Forests of Gold*, 58, cites J. Dupuis, *Journal of a Residence in Ashantee* (London, 1924), 65–66.

11. Wilks, *Forests of Gold*, 58.

12. Ibid., 66–72.

13. See especially, Klein, "Akan," 248–73.

14. Kojo Amanor, *The New Frontier. Farmer's Response to Land Degradation: A West African Study* (London and Geneva, 1994), 173–78. For the arguments that this is a historical process and not merely recent adaptation see, Fairhead and Leach, *Misreading the Landscape*, 289; Fairhead and Leach, *Reframing Deforestation*; and Wilks, *Forests of Gold*, 56–63.

15. Amanor, *New Frontier*, 175.

16. Ibid., 177.

17. Wilks, "Land, labour, capital," 494–95. Below I have cited many of the same primary sources he has used.

18. T. E. Bowdich, *Mission from Cape Coast to Ashantee* (London, 1819), 31, 325.

19. Wilks, *Forests of Gold*, 47.

20. Dupuis, *Journal*, 69, cited in Wilks, *Forests of Gold*, 49.

21. T. E. Bowdich, *The British and French Expedition to Teembo* (Paris, 1821), 18, cited in Wilks, *Forests of Gold*, 50.

22. For the advantages of grain over tubers in state formation, see McCann, *People of the Plow*, 153–4.

23. For this information on maize varieties, I am grateful to Peter Sallah and S. Twumasi-Afiyie, maize breeders at the Crop Research Institute in Kumasi. Interview 1/7/97. For cocoyam data, see Wilks, *Forests of Gold*, 52, and McCaskie, *State and Society*, 27.

24. Wilks, *Forests of Gold*, 84 n. 51.

25. Amanor, *New Frontier*, 33.

26. Polly Hill, *The Migrant Cocoa Farmers of Southern Ghana: A Study in Rural Capitalism* (Cambridge, 1963), 219–34. She cites a Basel missionary record book, stating that cocoa replaced dense virgin forest.

27. For the classic study, see Hill, *Migrant Cocoa Farmers*.

28. Fairhead and Leach, *Reframing Deforestation*, 67 (page citation is from draft manuscript).

29. Amanor, *New Frontier*, 64–66.

30. Kingdon, *Island Africa*, 107.

31. Ibid., 107.

32. Hill, *Migrant Cocoa Farmers*, 23–25.

33. Amanor, *New Frontier*, 64–65, describes the process of his Krobo study area and attributes it to an overall process of "degradation."

34. Marc P. E. Parren, "French and British Colonial Forest Policies: Past and Present Implications for Côte D'Ivoire and Ghana," Working Paper No. 188, History of Land Use Series, African Studies Center, Boston University, 1994.

35. World Bank, "The Forest Sector: A World Bank Policy Paper," 1991. Fairhead and Leach, *Reframing Deforestation*, cite the World Bank and many other alarmist figures.

36. The Kakum forest north of Cape Coast, for example, included open land reserved to protect the coastal settlements' scarce freshwater resources.

37. R. F. Soza, B. Asafo-Adjei, S. Twumasi-Afriyie, K. O. Adu-Tutu, and B. Boa-Amponsem, "Increasing Maize Productivity in Ghana through an Integrated Research Extension Approach," unpublished paper, Crop Research Institute, Ghana, 1996.

38. Centro Internationale por la Mejoridad de Maize y Trigo, "Ghana's Tradition Makers," *CIMMYT Today* 13 (1989): 2–10.

39. No-tillage cultivation utilizes seed drills to minimize soil disturbance, weed growth, and moisture loss from direct sunlight action on the soil. Personal Communication, Ben Dzah, Crop Research Center (Kumasi), June 1997.

40. R. F. Soza et al., "Increasing Maize Productivity," 1.

Chapter 7

Soil Matters: Erosion and Empire in Greater Lesotho, 1830–1990[1]

Twentieth-century travelers who have crossed the Caledon River from South Africa's Orange Free State and onto the lowlands and foothills west of Lesotho's Maluti mountains invariably remark on that landscape's most salient features: sandstone cliffs, Afro-Alpine highlands, and *dongas*.[2] Dongas are deeply eroded gullies and watercourses that stretch like fern fronds from the foothills downslope onto the lowlands, spreading branches that swallow footpaths, roads, and cultivated fields. Dongas even devour terraces, the recent human attempts to halt these gullies' inexorable spread deeper and further downslope. Viewed from a low-flying airplane, these expanding fissures in the landscape resemble spreading cracks in a badly shattered windshield. Moreover, dongas are not merely scars on the soil's surface; they exhibit different stages in a visible life cycle, varying in depth from 1 meter to 10 meters and in width from an easy jump for a person from edge to edge to a chasm capable of swallowing a house or a car. Even jaded development specialists accustomed to gully erosion in Ethiopia, Kenya, or the southwestern United States are stunned by the depth and the extent of Lesotho's dongas.

What is particularly remarkable about Lesotho's eroded gullies is, first, that they appear in their full development on the landscape almost exclusively on the Lesotho side of the border demarcation between Lesotho and South Africa's Free State (formerly the Orange Free State). Second, they appear quickly and grow in length, depth, and age like a living organism rather than a geological formation. Lo-

Figure 7.1   Donga near Roma, Lesotho, 1996. Photo by author.

cal people—sometimes the very young or the very old—recall a time
when a particular donga was either very small or when it was not
there at all. Some dongas visible in the 1990s are older and have lost
their vigor, others have been filled with rainwater, and still others have
stabilized under the cover of vegetation or have collapsed upon themselves, as if in old age. Younger ones nearby, however, may continue
to thrive and grow. Their age and life cycles seem to exist within
human rather than geological time scales.

As remarkable as they are in their speed of growth, their digestive
powers, and the layered colors of their strata, dongas take a toll on
their human hosts—the farmers and new urban dwellers of Lesotho—
by creating yawning space in what used to be a farmer's field, a road,
or a pasture. They are conduits for the loss of Lesotho's precious top
soil that travels downslope; during major floods this precious resource
even exits Lesotho, reaching South Africa via one of the region's major
river systems.

In the mid-1930s, Alan Pim, a British colonial official who was
surveying Lesotho's economic potential, identified soil erosion as "the
most immediately pressing" threat to the colony's economic future.
To address this crisis, Pim prescribed a draconian master plan of ter-

racing, diversion furrows, and meadow strips.[3] The length and impact
of this massive program transformed Lesotho's rural landscapes, perhaps as much as the donga themselves. Between 1935 and 1964,
207,872 hectares had been terraced with 42,747 kilometers of terraces;
2,555 kilometers of diversion furrows, and 2,254 kilometers of
meadow strips, all in a country the size of Belgium.[4] The colonial
assault on soil erosion in British Basutoland thus formed part of the
larger degradation narrative described for French West Africa, Kenya,
and postwar Ethiopia that associated the loss of natural resources with
local land use. In southern Africa, to colonial officials and travelers
who looked across the Basutoland frontier to South Africa's white
farms in the Free State, it seemed easy to conclude that Lesotho's
farmers had created dongas and degraded their own landscapes. To
add fuel to the fire, there is strong empirical evidence that productivity
on Lesotho's farms has suffered a steady decline since at least the
1950s, a seemingly obvious consequence of degraded soil resources.

The general decline in both total production and productivity has
continued in the 1980s and 1990s. During the 1980s the share of agriculture in the gross domestic product fluctuated between 20 percent
and 26 percent. By 1991 it had dropped to 13.9 percent. In 1976 the
portion of household income derived from agriculture had declined
to 18 percent. From 1974 to 1994 the mean annual production of cereal
crops per capita declined from 180 to about 70 kilograms.[5]

The case of Lesotho's dongas fits neatly into two radically opposed,
but oddly complementary, readings of African landscapes. In the first
instance the presence of eroded gullies on prime farmland in Lesotho
seems to offer prima facie evidence for the dual economy argument
of apartheid economists who argued that African farming techniques
were both unproductive and destructive of natural resources. Such a
view only had to glance at the minimal erosion on white farms around
the town of Ladybrand across the border from Lesotho in the Orange
Free State to confirm this conclusion. Yet, that same evidence and the
overall decline of Lesotho's farm economy reflected in Table 7.1
equally supports a mirror image of the dual economy argument.
Scholars of political economy who opposed apartheid were able to
argue that Lesotho's status as a classic labor reserve to South Africa's
mine economy labor sapped Lesotho of its economic viability. Chapter 1 of Colin Murray's classic study *Families Divided* poignantly tells
the story of Lesotho's fall "from Granary to labour reserve."[6] He cites
the evidence for Basotho engaging in agricultural entrepreneurship in

Table 7.1
Grain Production and Productivity in Lesotho, 1950–70

	Maize		Sorghum		Wheat	
Year	A*	B†	A	B	A	B
1950	214	11.9	49	8.7	50	10.1
1960	121	7.4	54	7.8	58	8.5
1970	67	5.2	57	6.9	58	5.4

A* = total annual production.
B† = Quintals per hectare.

Source: Colin Murray, *Families Divided: The Impact of Labour in Lesotho History* (Johannesburg, 1981), 19.

the 1870s when farmers in Basutoland (then under the authority of Cape Province) provided massive amounts of grain to feed the diamond mines at Kimberly (see below) only to succumb to the subsequent loss of their agricultural base through South African and British colonial efforts to tap Basotho as a source of labor rather than as a source of food for the mines. Murray's careful and sensitive research, essentially a social history, portrays the tragedy of South Africa's political economy, but glosses over the finer-tuned role of the historical conjunctures of soils, farmers, and landscapes. This chapter will fill in some of those details in a way that supports Murray's basic argument.

Kate Showers, a soil scientist with an active interest in history, proposes an alternate approach that directly seeks an explanation for the spider webs of dongas that cover much of modern Lesotho. Using Basotho oral evidence as well as published records, Showers traces the rural tragedy of soil erosion to colonial intervention itself, arguing that erosion existed on a relatively small scale until the 1936 anti-erosion campaign imposed its massive intervention of terraces, meadow strips, and furrows on Lesotho's countryside. Her argument extricates farmers from the equation in favor of a primary causal role for poor engineering and the coercive tactics of the colonial project. Showers argues that farmers whom she interviewed in a fairly remote rural research site recalled donga as a recent phenomenon (i.e., within their lifetime) rather than as a historical feature of the landscape. In Showers's view the colonial government's efforts ironically forced farmers to build the terraces and meadow strips that ended up concentrating the corrosive effects of water and magnifying its effects. In her account, farmers in the Free State, by contrast, faced no such coercion and thus have not suffered from such massive erosion.[7] Bad local management was at fault; it was not the Basotho farmers themselves but the cost-cutting coercive power and poor planning of a myopic colonial government that accounted for the degradation.

But are there wider geological, historical, and human factors that account for Lesotho's scarred landscape and its declining agricultural productivity? My argument here sees gully erosion as a geological phenomenon with a social and economic history that predates conservation interventions of the 1930s. This chapter places the evidence of Lesotho's soil erosion into a full context of environmental history: the presence of dongas on Lesotho's landscape is a result of cumulative factors of physical geography, human strategies of production, and the historical conjuncture of these local issues with southern Africa's modern political economy. While no single factor can fully account for dongas' place in Lesotho's rural scenery in the late twentieth century, the peculiarities of its soil and its agricultural history—as opposed to pure issues of political economy or colonial rule—take pride of place. Nonetheless, the historical interaction of human activity and the physical world suggests that the origin of dongas necessarily implies a movement through time that challenges the assumptions of most arguments about degradation.

HIGHVELD AND HIGH GROUND: GREATER LESOTHO'S PHYSICAL BASE

Lesotho as a nation-state and as a specific ecological setting took shape as a historical byproduct of the geophysical political, military, and environmental forces that propelled southern Africa as a whole. Greater Lesotho (the land of the Basotho people) lies in a particular configuration of southern Africa's southern highveld that provided key mountain outcroppings for military defense and soils upon which the Basotho people fashioned a productive agricultural economy in the middle and late decades of the nineteenth century. The historical homeland of the Basotho lies in the zone of grassland and uplands between the Vaal River valley in the north, the Orange River valley in the south. It includes the Maluti mountains (called the Drakensburg in South Africa) in the east and lowlands in the west that stretch from the foothills of the Maluti westward toward Thaba Nchu (Black

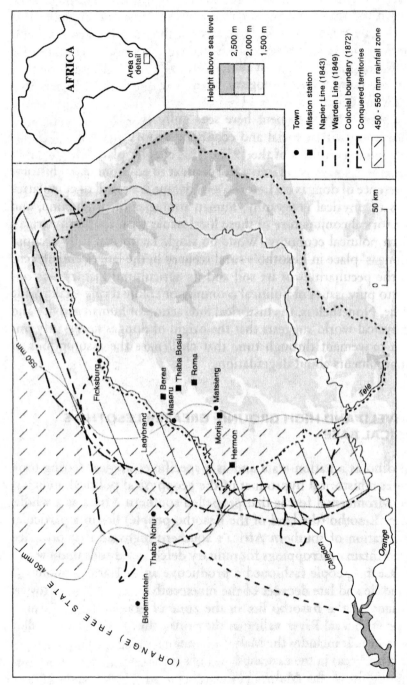

Figure 7.2 Map of Lesotho/Free State.

Mountain), one of the many cave sandstone mesas that dot the low-land plain as it rises slowly east toward the mountains on the horizon. These flat-topped outcroppings provided shelters for late Iron Age people who left behind the stone settlement enclosures that served as the stage for much of the nineteenth-century struggle over land and ideas about landscape. Greater Lesotho marks its western frontier less from geology than from the imaginary line defined by the 500-millimeter annual rainfall isohyet. This barrier is key to human geography since annual moisture less than 500 millimeters prevents the cultivation of maize and wheat, two crops central to this historical drama of soil, politics, and environment.

The defining geological process for Lesotho's modern history is the successive layering of geological materials heaped upon the Karoo (desert) sandstone base. The volcanic activity that created the Maluti Mountains laid hard basalt and dolorite formations on a peculiarly soft base of sandstone, shale, and mudstone. These latter materials are argillaceous (clay) or sandy rocks that quickly break apart when exposed to the elements of nature.[8] The slow, corrosive action of rainfall, wind, and water over the basalt and dolorite that overlay the soft underlayer gradually deposited distinctive layers of volcanic loam on the lowlands to the west. The sandstone cliffs and mesas mentioned above are the remnants of the Karoo (desert) sediment that remained beneath the dissolved basalt. These odd mesa formations—Thaba Bosiu, Thaba Nchu, and so forth—have become synonymous with Lesotho's historical inhabitants: they are the sites of dinosaur footprints and Stone Age rock painting; for Basotho people they also capture romantic imagery of state formation in the nineteenth century. In addition to flat-topped sandstone formations, the dolerite sills and the dykes that appear as outcroppings at odd angles are the other major feature of these lowlands. These outcroppings take their form because their igneous rock material is harder than the basalt and sandstone layers that decomposed more readily around them. In the twentieth century the sight of these distinctive mesas on the horizon has come to symbolize to Basotho migrants returning from South Africa their arrival home from the life of the mines and the city.[9]

The soils that formed in the Caledon River valley from this geological history have a distinctive personality. They are predominantly "duplex" soils that contain a top layer of sodic loam, are low in humus, but are very friable because of their high sand content. Beneath this top level is a layer of clay soils that resists penetration by water.

These soil layers behave in peculiar ways when they come in contact with water. In low-lying basins they take the form of *vleis* (marshy areas or marl bogs) that retain surface moisture well past the summer rainy season. In past times Basotho hunters used these mucky bogs to entrap game. Such formations also have strong symbolic significance for Basotho since it was one of these wetlands, Ntshoana Tsatsi (in present-day Free State), where they believe animals and humankind emerged simultaneously from a hole in the ground (see below). This symbolism contrasts with Christian teachings about the Garden of Eden and humankind's role as master of the natural world. After 1840 when Basotho moved down off the defensible high terraces to settle on bottomlands, this moisture-retaining characteristic allowed innovative Basotho farmers to adapt quickly to cultivation of winter wheat that needed late-season moisture for germination.[10]

The second, less agreeable, feature of these duplex soils is their response to water movement along stream banks and heavily grazed pasture or on plowed agricultural fields. Under moisture the high sodium content of the upper soil layer causes it to dissolve while the lower clay stratum seals itself, concentrating the force of the flowing water on the top layers, resulting in erosion in the form of gullies. Gullies, or dongas, thus have formed in areas where the force of moving water concentrated, that is, on slopes below mesas or where fire removed vegetation and exposed top soil to a brisk movement of surface water. Beneath the surface as well, these soils have also been susceptible to "piping," where clay layers channel water underground, eventually causing the collapse of the surface layer and creation of a young donga or a new tentacle of an older gully. In human terms these soil characteristics have therefore been a two-edged sword: they create conditions for retaining soil moisture at the same time that they expose the soil to a particularly dramatic form of erosion. The dominant soils of Greater Lesotho are thus extraordinarily sensitive to surface disturbance and concentrated movements of water of any sort. Human events of the nineteenth century played themselves out on this fragile and movable stage.

That dongas existed as a part of the natural landscape before modern settlement seems clear from both historical and geological evidence. Earliest written descriptions of travel in Greater Lesotho depict existing stream beds and ravines as deeply cut into the soil. Eugene Casalis, a Protestant Evangelical Mission Society (PEMS) missionary who arrived in Lesotho in 1833, described his group's reaching the

edge of a "steep ravine" that barred their way. That this barrier was
not a riverbed or a rocky ravine but a donga carved from soft loam
and clay is clear in that they burrowed their way through the em-
bankment without any iron tools at hand, "now scraping the ground
with our bare hands, now using a sharp stone for want of a spade."[11]
In 1833 such formations appeared to travelers to be a natural, if an-
noying, part of the natural landscape rather than a sign of degradation
or a threat to human welfare. Other events taking shape in the region,
however, created a new set of physical forces that by the latter quarter
of the nineteenth century would place human hands and tools on the
land and account for a rapidly expanding network of dongas through
the lowland soils.

GREATER LESOTHO AS A STAGE FOR STRUGGLE[12]

In the first decades of the nineteenth century, a series of environmental shocks in the area of the Orange and Vaal Rivers caused a
crisis of both subsistence (failed grain harvests) and social reproduction (loss of livestock to drought and epizootic disease) that tore open
the region's social fabric and human geography. The resolution of
these human crises set the stage for the region's subsequent landscape
history. From the Cape the leading edge of European settlement began
an expansion of its livestock frontier with cattle raiding, an active slave
trade, and the marauding of armed, horsed groups of Griqua, Europeans, and Khoisan.[13] From the southeast from across and around the
Drakensburg Mountains, the violent Zulu expansion (*mfecane*)
pushed waves of refugees before it. Those societies that were less powerful technologically or less well organized militarily were crushed,
scattered, or forcibly absorbed by these two forces. In the search for
security, subsistence, and defensible turf, elements of Sesotho-
speaking lineages and other bits and pieces of refugee groups sought
refuge under the leadership of a young Basotho chief named Moshoeshoe.

In a cold winter trek in 1824, Moshoeshoe led his people and livestock out of harm's way off the open lowlands and onto a barren
mesa, Thaba Bosiu, a day's walk east of the Caledon River valley.
Thaba Bosiu was ideal defensive ground: it commanded the mountain
pastures behind it as well as the region's best watered lowlands before
it to the south and east. Best of all, the narrow passage to the top of

the mountain allowed the Basotho to protect their herds from better-armed Zulu regiments and Cape cattle raiders on horseback.

Moshoeshoe created security for his constituency through military leadership and his canny talents in using his cattle as social capital. He forged a Basotho nation with social and political coherence. At the same time, he set out to define a geographic and ecological base for the nation. Though the term *Lesotho* implies a political/territorial unit, the evidence indicates that Moshoeshoe initially thought less in terms of fixed boundaries than in terms of the people who acknowledged his authority and of landscapes with specific cultural memories (see below).[14] Nevertheless, from Thaba Bosiu, Moshoeshoe could claim and defend territory that included the fertile foothills in the west and, at his back, the Afro-Alpine ecology of the Maluti Mountains that provided summer grazing on sweetveld grasses at highland cattle posts. These posts provided both security from raiding as well as isolation from disease for local Afrikander cattle, fat-tailed sheep, and goats. If the highlands offered few trees—only occasional willows and wild olive (*Olea africana*) and woody shrubbed ravines—the Afro-Alpine ecology offered abundant pasture. Livestock, particularly the cattle, provided a social bank account from which Moshoeshoe endowed marriage alliances to cement and expand his chiefdom.

As security conditions improved in the 1830s, lowlands to the west and southwest offered *Themeda triandra* grassland ideal for winter grazing; the landscape also presented natural alluvial terraces enriched by the black and brown loam washed down from the basalt and dolorite of the mountains above and mixed with the lowland sandy and clay soils. Moshoeshoe could claim control of the strip of 10 to 40 kilometers of this land up to the Caledon Valley and perhaps a-day-and-a-half's walk beyond on the well-watered plains. This largely treeless lowland plain required little labor for clearing vegetation and its friable topsoil yielded easily to Basotho women's hand hoes. Basotho cultivated crops of sorghum and maize that sustained the Basotho diet of sour milk, sorghum bread, and porridge and the chance to offer grain in exchange to neighboring groups from more-arid pastoral zones. It was not an accident that Greater Lesotho's borders on the west included an environmental feature crucial to the landscape's economic value: the 500-millimeter rainfall isohyet—the minimum annual moisture requirement for maize.

The basaltic mountains and sandstone cliffs that provided Lesotho's defenses within modern human history had in geological time laid down a fragile treasure in the form of deep soil sediments in river valleys and at the base of its cliffs. This setting of highlands, palatable grasses, and deep, friable soils underwrote the economic base of Moshoeshoe's new society. Yet, the landscape was an untested one for the Basotho farming system based on the hand hoe and was *terra incognita* for the new requirements of the crops and tools that arrived as a part of southern Africa's emerging political economy.

Initially, the farming system brought by the Basotho from their homeland in the Vaal River basin adapted well to the new possibilities. By the early 1830s, settlements that had hidden themselves in the protected hills had descended cautiously to form villages alongside fields of maize and sorghum that women prepared with hand hoes and weeded cooperatively. In this era, sorghum was the preferred food and the dominant crop. The friable, sodic soils required little clearing and plot preparation and allowed Basotho to escape to defensive positions in the foothills at any sign of an enemy. Eugene Casalis, a French Protestant missionary, described in his diary his first encounter with Basotho at the western edge of their territory in June 1833:

> Two days later we reached the foot of a beautiful mountain [Thaba Nchu] five to six hundred metres in height and several kilometers in circumference. Directly under the mountain, we could see large fields of near ripe maize and sorghum (large millet). The inhabitants had built their huts on the steepest summits, as a precaution against enemy attacks. Those who were busy in the plantations fled at our approach.[15]

By the 1830s, the Basotho had reclaimed the lowland territory that included winter lowland grazing on *Themeda trianda* pasture and "abundant" water sources. Under Moshoeshoe's Bakwena lineage (the crocodile clan), a tributary land tenure system evolved under the control of local chiefs that allowed each household to claim three plots to take advantage of soils types, microclimates, and crop rotation. Livestock grazed communally on grasses largely untouched over at least a generation. The conjuncture of soils, labor, and market led the Basotho to the threshold of an agricultural revolution that would define their new landscape.

To this human landscape, reshuffled and redealt by the Mfecane, came a further element, this time from the southwest where a seventeenth-century implantation of European peoples had taken root at the Cape of Good Hope. By the late 1830s, the frontier of marauding Griqua and Kora raiders had been replaced by Voortrekkers, Cape Dutch frontier folk, who sought to settle the highveld to escape British rule at the Cape. As pastoralists who kept in trade contact with the Cape economy but who sought an independent economic base in the Vaal River valley, the Voortrekkers initially sought sources of water for homestead sites and viable pasture for their cattle and sheep rather than agricultural land. By 1837 they had pushed the Ndebele out of the Transvaal and in 1838 decisively defeated the Zulu under Shaka's successor Dingane at the Battle of Blood River. The trickle of settlers who arrived at the edge of Lesotho's territory near Winburg after 1836, however, behaved like other new settlers, and Moshoeshoe extended them welcome under similar terms of settlement as he had offered other groups who were seeking access to land.[16]

The new European arrivals, however, were only the first wave. Over the next decade their numbers increased, and in 1846 they settled at a site of a local freshwater spring named Bloemfontein ("fountain of flowers"), close to the territory at Thaba Nchu that Moshoeshoe understood to be his own. The deadly struggle over the meaning of land claims and over conflicting visions of landscapes in the next decades not only defined political boundaries but also, ultimately, fashioned Greater Lesotho's human and physical landscapes.

These elements of environmental history took shape, however, as the product of a competing set of ideas about natural resources and the order of nature. In 1833 Moshoeshoe sniffed the political winds and dispatched a Christian convert Griqua, named Adam Krotz, with 200 cattle "to procure in exchange at least one missionary."[17] In fact, Moshoeshoe got three for the price of one: that same year a trio of French Protestant missionaries from the Paris Evangelical Missionary Society (PEMS) arrived and immediately sought a site for planting the seeds of Christianity: "After a prolonged search, we fixed upon a spot which seemed to offer every possible advantage, water in abundance, a fertile soil, firewood, timber, and a picturesque situation."[18] The place they selected was a well-watered and (unusually) wooded valley they called Morija. Their choice of the site reflected the ideals of nineteenth-century European naturalism as well as a plan to germinate Christianity by building a wood-frame mission settlement and cultivating fruit trees, wheat, and other exotic cultigens year-round.

The Basotho greeted the Christian message emanating from Morija with cautious interest but absorbed and adapted with alacrity ideas about nature, land management, and vegetation. In succeeding years the arrival of other mission groups of Roman Catholics (1864 at Roma) and Anglicans (1876 at Masite) provided the Basotho with a set of key interlocutors for new technologies, exotic crops, and new ideas about ideal habitats. The political and military conflicts that settled the political boundaries of southern Africa in the second half of the nineteenth century did more than lay the foundations for the apartheid state (1948–94); struggles over borders were really struggles over nature (natural resources) and over visions of the land.

LANDSCAPES OF MEANING, LANDSCAPES OF SOIL

It is difficult to know what ideas about land and landscapes that the Basotho who were living around Thaba Bosiu in the 1830s might have had and debated. The trauma of invasion, loss of loved ones, allegations of cannibalism, and retreat to the biting winter cold on the "Mountain of Night" may in fact have rendered such debates esoteric. Food and security must have been uppermost in their thoughts. Men likely thought of warfare and defense while women felt most concerned about the dangers they faced while cultivating their fields on the bottomlands in the valley below the protected mountain top. Nature must have seemed to offer more threats than opportunities. Hilltops and steep ravines meant security; fields and pasture on the open valleys were unsafe, contested ground.

The Sotho theologian Gabriel Setiloane tells us that the Basotho describe nature and humanity as having emerged simultaneously at creation from an opening in the earth. God (*Modimo*) dwelt within the earth (*Mosima*) and thus created humans and animals as part of a fully formed natural world: "Our first parents came out of a hole in the ground. They came out together, men with their wives, children and their animals, cattle, sheep, goats, and dogs."[19] Human landscapes thus necessarily included households and domestic livestock, but not a specifically defined territorial setting. The Bakwena clan (Moshoeshoe's own lineage) claims that its arrival on earth took place in a marshy bed of reeds at Ntsoana Tsatsi, in the east of the present Free State from which Moshoeshoe's clan migrated to greater safety in 1824. Thus, while Moshoeshoe's people arrived on earth with their livestock, there was no Eden, no place granted to them by God in a way described to them by missionaries.

Yet, landscapes did have strong cultural meaning. The peoples who occupied land east and west of the Caledon Valley had distinctive conceptual frameworks for imagining the land. European observers who traveled with Basotho into new areas marveled at their ability to use small pieces of landscape as mnemonic mapping devices to read and recall spatial orientation.[20] They never seemed to lose their way and ascribed meaning to landscapes that Europeans found quite remarkable. Moshoeshoe, for example, offered the missionary Thomas Arbousset a nostalgic tour de force of the cultural meaning of northern Lesotho landscapes as they traveled past the terrain of the king's childhood, ascribing personal meaning to specific rocks, trails, and water sources. His associations were far more social and symbolic than economic, more painful than acquisitive. Arbousset's diary recorded Moshoeshoe's narration:

> Over there, rises the mountain where Mokhachane was circumcised. To the right of that, you see the place where he lost a famous battle. Look at these remains of ancient kraals opposite us: they belonged to the wise Matete. What you are seeing in the same direction, a little farther off, are the ruins of the village of Mahao. Oh, how many human skulls lie in the depths of these valleys! How sad they seem! In olden days these mountains were alive with people; the cattle delighted in going out to pasture on those heights over there; all those hills facing us were cultivated . . . nowadays I see nothing but death, I hear nothing but silence.[21]

Basotho ideas about nature and the ideal setting of humans in nature probably changed dramatically over the nineteenth century as politics, new ideologies of identity and nationhood, Basotho identity, Christianity, and new economic forms swirled around them. Historian Elizabeth Eldredge has characterized the dominant Sotho idea in this period as "the pursuit of security."[22] In the 1820s and 1830s no doubt the ideal Sotho human landscape was a village huddled against a sandstone cliff, secure from attack, with fields of sorghum and maize and protected pasture for livestock. Agriculture was female: a secure household food supply rather than an economic enterprise. Cattle were male social capital; sheep and goats were small change and food to call upon in hard times or for ritual events.

For his part, Moshoeshoe's narrative indicates that he saw idealized nature not in terms of bounded land and vegetation but as people, villages linked by lineage and tributary obligations to him and to one

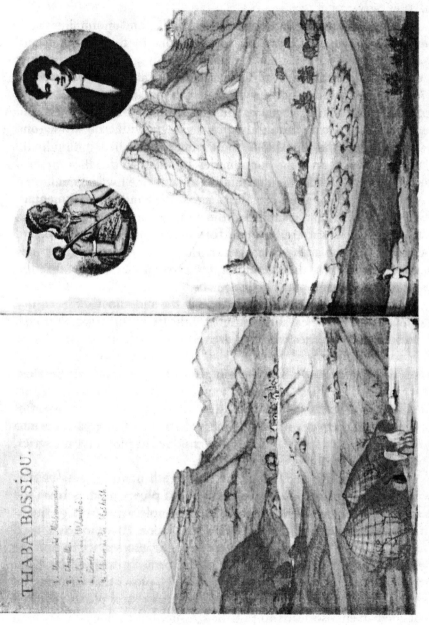

Figure 7.3 Settlement at Thaba Bosiu. Painting by Ina-Maria Harris based on an earlier drawing from the nineteenth century. (Upper right, portraits of Moshoeshoe and missionary Thomas Arbousset, c. 1840). From Thomas Arbousset, *Missionary Excursion* (Morija: Morija Archives, 1991).

another. This view of security and the natural world would have excluded abstract notions of nature devoid of human claims or seen as partible property.

The Lesotho setting of lowlands, foothills, and mountains was a new one for the newly formed Basotho nation; by the mid–nineteenth century it was still less than a generation old, a landscape that met their view of security and subsistence as an ideal. The barren, treeless plateaus and foothills that early travelers complained about as "disagreeable" and lacking scenery represented to the Basotho safety and an appropriate setting for tightly drawn villages with crop storage and livestock.[23] By the late 1830s these conceptions had undoubtedly changed. Newly arriving missionaries found that the Basotho had moved down from the high ground to inhabit the lowlands in homesteads scattered like grain broadcast across the landscape. Segmentary lineages hived off and established themselves on new land; new lowland settlements scattered around former pasture. Moshoeshoe, and others of Lesotho's founding generation, however, viewed this new human geography as disquieting, and they opposed permanent settlement of the lowland plain as dangerous.[24]

A second view of Lesotho landscapes in the mid–nineteenth century dominates the historical sources. In 1833 the first generation of French Protestant missionaries who had arrived to establish permanent mission posts brought with them not only a firm view about the spiritual needs of Moshoeshoe's Basotho but also a zeal to transform the treeless grassland of their "African Pyranees" into a wooded landscape of Christian husbandmen. Along with the gospel, missionaries brought with them new tools and ideas for transforming pastures into fields and meadows into orchards: the moldboard plow, tree nurseries, and wood frame domestic architecture.

Over the middle decades of the nineteenth century, missionaries were the first to introduce steel moldboard plows; traders from the Cape and, later, Kimberly imported these implements in large quantities in exchange for livestock, grain, and labor. Basotho oxen, used traditionally for carrying loads, were brought into service to pull the heavy plows for deep plowing of fields formerly worked by hand hoes. Unlike light, wooden Ethiopian scratch plows that farmers carried on their backs to the field, Basotho heavy steel plows required that their oxen also learn to pull sledges needed to drag the tools to and from the homestead and to carry grain from the threshing floor to storage and thence to market. The new plows and Basotho skills in training oxen allowed a dramatic expansion of cultivated land and responded to new markets for grain with the opening up of new mining centers, first at Kimberly (diamonds in 1867) and later on the Witwatersrand (gold in 1886) and still later in the Free State itself. Other Basotho earned a living by using teams of oxen to pull wagon loads of valuable grain to Free State markets.

Missionaries also brought new ideas about the need for trees for fuel and building material and for their aesthetic value as markers of human settlement. Letters home from the first missions repeated images of barren mountains and successive meadows without shade and fuel. Once established at mission stations, Europeans began the task of civilization, that is, building churches, schools, and workshops. For this they needed wood beams and poles. Thus trees and plows spread in tandem, etching new colors and textures on Lesotho's landscape itself. The Basotho saw missionaries as part and parcel of the Christian work of the spreading gospel, though it is likely that more Basotho accepted Christian ideas about landscapes than they did Christian ideas about the hereafter.

While the Basotho came to appreciate trees and became enthusiastic afforesters, especially of fruit trees, their mid–nineteenth-century methods of land management and human landscapes were antithetical to tree cover. Trees located near sorghum fields gave convenient roosts to pigeons and other birds that fed on ripening grain.[25] Moreover, clearing fields for cultivation or for new growth of sweetveld grasses called for fire; burning grassland tended to eliminate most woody vegetation except fire-resistant wood scrub species (*sehalahala*). Most trees therefore nestled in stream beds or in mountain ravines safe from grass fires.[26] In the mid-1830s the major wood resources were willows, wild olive (*Olea africana*), and woody shrubs. These quickly fell to the Christian ax. Missionaries who arrived in the 1880s, however, lamented the destruction of those hardwood species for the sake of God's work (building churches). In December 1850 the PEMS missionary François Maeder professed disquiet about the higher calling of the mission's need for timber:

> I have often admired the wild olive tree, with its small dark leaves and its far-spreading branches. It is fond of growing in the crevasses of our mountains. . . . The beauty of these trees fascinates and holds the attention when one is all too familiar with the sight of semi-arid mountains with no other covering than a sprinkling of grass. And yet, after en-

joying the cool shade of these groves and admiring their beauty, I have
no choice but to apply the destructive axe to the tree trunk and to bring
it down; for necessity converts the act into a duty.[27]

Missionary axes and nurseries changed Lesotho's tree cover. Poplar,
eucalyptus, oak, blue spruce, peach replaced native Afro-Alpine species. Rather than mark remote, inaccessible spaces, trees in Lesotho,
as in highland Ethiopia, came to denote settled space. The mission
settlements at Morija, Roma, and Berea and then later at urban centers
at Maseru and Mafeteng appeared on the landscape as green, semi-
forested spaces soon after their founding. Moreover, the Basotho
abandoned their squat houses built from saplings and reeds (*mohlon-
goa fatsha*) in favor of circular thatched houses (Sesotho: *rontabole*)
made of eucalyptus poles. In this effort Moshoeshoe led the way,
building a Dutch-style, rectangular gabled house at the pinnacle of
Thaba Bosiu.

Thus, both the Lesotho landscape and Basotho ideas about the appropriate look of rural scenery changed slowly but substantially. For
his part, Moshoeshoe, with the advice and counsel of the missionaries,
came to appreciate the boundaries anchored in written title to the land
and not merely expressed in human relations of dependence. These
were hard-won lessons learned in the conflict with European—Voortrekker and English—images of frontier landscapes fought in the middle decades of the nineteenth century. These battles over political
borders were direct claims on resources of pasture, water, and arable
land, but also struggles over visions of landscapes.

Voortrekkers had a very different view of the land and resources
between the Vaal and the Orange Rivers that they surveyed in the
1830s and 1840s. Their visions diverged from the Basotho in two fundamental ways. First, unlike the Basotho, who were mixed cultivators
who took their cattle very seriously, Voortrekkers were dedicated
pastoralists who saw the Themeda trianda grasslands as open grazing
for cattle and sheep destined for markets at the Cape. Second, Voortrekkers displayed the rough-and-tumble attitude of frontier people
who didn't mind raiding outsiders' livestock while nonetheless em-
bracing a European idea of land ownership. Moreover, Moshoeshoe
appeared to them to control the best of the land, that of the lowlands
of the Caledon valley. Thus, the struggle over the landscape that ensued involved visions of ownership—visions of whose cultural im-

Figure 7.4 The settlement of Bloemfontein c. 1851 (Thomas Baines, artist). (Courtesy of Oliewenhuis Art Museum, Bloemfontein.)

print would expand—and the status of the people who occupied the middle ground.

The most significant imprint on the land was the 1846 founding of the town of Bloemfontein. By 1851 the settlement already showed clear signs of European settlement and cultural implantations—a church, government offices, and streets laid out in a grid pattern. Thomas Baines's 1851 painting showed new settlers etching their mark on the treeless landscape of the highveld.

SETTING BOUNDARIES, 1843–1884

After 1840 there was a steady stream of Boer farmers who were arriving from the Cape to join the Voortrekkers at the eastern edges of the Basotho territory where pasture and rainfall improved as one proceeded east. Boer frontiersmen and the Basotho traded cattle raids, eventually prompting Moshoeshoe to request intervention by the British government at the Cape. In 1843 the British governor George Napier imposed a treaty that set a formal territorial boundary (see map).[28] In 1854 after pulling back from the brink of several military confrontations with Moshoeshoe, the British government recognized the existence of the Voortrekker state as the Orange River Republic in the Treaty of Bloemfontein, abrogating the agreement of 1843 with Lesotho. In 1858 formal war finally broke out over claims to land by settlers, pushing lines of Boer settlement steadily east onto the Basotho's better-watered agricultural lands. While the British declared themselves neutral (but supplying arms to the Free State), the Boer army successfully raided and burned several mission stations including the PEMS settlement at Morija. The peace treaty brokered by Sir George Grey, governor of Cape Colony, set new boundaries but maintained an arms embargo on Lesotho. Moshoeshoe correctly perceived the real object of the Boer hostility and appealed to British authority for protection.

In 1866 war broke out again and went badly for the Basotho who lost most of the lowlands to the Boer army when Chief Molapo, Moshoeshoe's own son, ceded his prime agricultural land to the Free State. Moshoeshoe again appealed to the British for protection, and in January 1868 the British government in London agreed to annex Lesotho. In the face of the British intervention, the Free State agreed in 1869 at the Treaty of Aliwal North to restore the boundaries of 1858. While the Free State was never able to complete its conquest of Lesotho, they kept control of what the Basotho still call the "conquered territories," land west of the Caledon controlled by the Basotho since the 1830s. Administered first by Natal and then by the Cape, in 1884, after the disastrous (for the British) Gun War of 1880–81, Basutoland became a Crown Colony with its current borders.[29]

The political and military wrangles over borders masked fundamental changes in land use and population that were taking place in Lesotho. Trade contacts continued to open and expand markets for grain shipped by oxwagon to growing urban centers like Kimberly and Bloemfontein. This trade also brought plows, new crops, Merino sheep, and Angora goats, all defining elements of Lesotho's twentieth-century countryside. Confined within a bounded ecological zone and cut off from the natural extension of its lowlands, Lesotho nonetheless joined the regional economy of the Free State and the industrial revolution in South Africa.

The discovery of diamonds at Kimberly in 1867 transformed the regional economy of Basutoland and the Orange Free State. Two of the first impacts of the influx of workers and capital to the north-eastern Cape was a market for food and the birth of an economic infrastructure in the region including a cash economy, road networks, and technology. A Morija missionary in 1871 noted:

> Hitherto our Basuto have all remained quietly at home, and the movement which is taking place beyond their frontiers has produced no other effect than to increase the export of wheat and other cereals to a most remarkable degree. While the district in which the diamonds are found is of desperate aridity, the valleys of Basutoland, composed as they are of a deep layer of vegetable mould, watered by numerous streams and favoured with regular rains in the good season, require little more that a modicum of work to cover themselves with the richest crops.[30]

The response of Basotho farmers was nothing short of astonishing. In 1873 alone Basutoland exported 100,000 bags of grain (about 8,400 metric tons) and 2,000 bags of wool. Export figures in 1893 had more than doubled to include 11,600 tons of wheat and 6,000 tons of maize and 500 metric tons of wool.[31] By 1900 the region's economy displayed the Basotho response to the new markets and the region's links to international markets in specific ways: wheat, maize, Merino sheep, and Angora goats superseded Basutoland's agricultural staples of sorghum, milk, and cattle.

Such rapid and profound economic changes by 1900 were already evident on the landscape and in the disturbance of the region's productive but fragile pediment soils. Imported varieties of small ruminants (sheep and goats) thrived on the native grasses but their concentration on the best pasture gradually encouraged the succession of less-palatable grasses. Cultivation of winter wheat took place around the new mission stations that introduced plow agriculture on land with residual moisture most sensitive to disturbance. More subtly, maize began to replace sorghum as the dominant national crop and daily food source. The sledges and oxcarts used to transport grain and equipment along cattle tracks further reduced ground cover, leaving Basutoland's soils little protection from torrential convectional rains, fire, and human traffic.

It is not surprising that early examples of major gully erosion arose from this era and were most evident around the sites of mission stations and concentrated plow agriculture. In the early 1890s Alice Balfour visited the French Protestant Mission station near Berea where

Figure 7.5   Cultivation of maize (with donga) at Morija, c. 1915. (Courtesy of Morija Museum and Archives.)

she commented on the presence of dongas created by the action of water pouring over the steep sandstone cliffs onto lowland soils:

> Immediately on reaching the plains the streams thus formed make *dongas* or watercourses through the soil which is here often of prodigious thickness. These dongas rapidly increase in size and depth as they go along, their sides being almost perpendicular, only furrowed with rain, which sometimes leaves the most fantastic pinnacles and spires of somewhat harder soil sticking up here and there. No doubt the formation of dongas, which intersect the plains in every direction, and which are so rapidly increasing in size and number, is greatly aided by the absolute absence of trees.[32]

Balfour's astonishment at the size and number of dongas at Berea points to their presence even before 1900. Yet her observations also suggest that gully erosion was not a feature of the landscape from time immemorial. At Berea she noted the evidence of the growth of dongas within the period after the arrival of missionaries and the plow:

> We were taken to see a donga close by. It differs in no wise from other dongas, except in being somewhat bigger than most. But the interesting

Figure 7.6 Donga at Berea, 1895. From Alice Balfour, *Twelve Hundred Miles in a Waggon* (London, 1895).

point about it is that its commencement as a little ditch across which you could jump, was seen by a missionary who only lately left here, and the whole of its growth is the work of fifty years. It is now, we guessed about 80 feet deep and 150 feet wide.

Finally, she noted the connection between disturbed topsoil, dongas, and areas of concentrated settlement:

Another day we drove to the Roman Catholic Mission Station called Roma, some twenty miles from Maseru. The Government spends so much time and money keeping the Basutoland roads in order, but no newcomer unacquainted with the washing out powers of the rain in this country would guess it to look at them. The road got worse as we went on, the dongas getting deeper and more frequent as we got to more hilly country.[33]

Balfour's observations corroborate Basotho accounts from ha Maghopo, a district remote from the road network, related by Kate Showers in the early 1980s. Her informants remembered a remote local landscape in the 1940s before dongas. Alan Pim's alarming observations of erosion in 1935 indicated that the land most affected was concentrated around settlement sites, roads, and paths (i.e., disturbed soils) in the lowlands.[34] Here the fact of erosion concentrated around commercial arteries is not so much an example of Robert Chambers's "bias of the road" as it is evidence that dongas accompanied any form of soil disturbance.

The rural landscape that Alan Pim described in 1935 was in fact a product of major conjunctural change as Basutoland fell into the maelstrom of southern Africa's modern political economy. The same regional economic boom that fueled Lesotho's agriculture also increasingly drew Basotho male labor to the mines and away from full-time agriculture. In 1873, 15,000 Basotho worked in South Africa's diamond mines; in 1886 (the year of gold's discovery on the Witwatersrand) the number had doubled. During the 1920s only one fifth of Basutoland's adult males were still employed in agriculture. Wage labor thus had rapidly supplanted agriculture as a source of cash.[35]

At the same time, events in the South Africa across the border forced Basutoland from an agricultural economy to a labor reserve. In 1893 the Orange Free State government imposed import tariffs on food grown in Basutoland in order to support its struggling white farms. Finally, the arrival of the railroad at Kimberly also opened the regional markets to cheap grain imports from Australia and the United States. The Basotho who had taken up sharecropping across the border in the Orange Free State faced increased landlord demands; those who had originally kept half of their crop by the turn of the century were able to retain only one third. The 1913 Native Lands Act resulted in mass movements back across the border into the more densely settled Basutoland lowlands.

Basutoland's entrepreneurial economy that had shown such resilience in the 1870s suffered two simultaneous shifts. First, the loss of male labor to the mines left local agriculture in the hands of women who assumed the role of managing the farms without male labor and who were forced to maintain food supplies. While men often had mine contracts that allowed them to return home during the agricultural season, in practice male labor was usually missing at the key work days for plow agriculture.[36] Second, the feminization of agriculture shifted strategies away from high-labor, uncompetitive wheat (available more cheaply from overseas markets) to maize. Maize was a high-yield, low-labor crop attractive to labor-poor farms. It was no accident that by 1930 maize had surpassed wheat as Basutoland's major cash crop and in turn supplanted sorghum as the major food crop. By the mid-1930s the colony had become a net importer of grain.[37] By the late 1970s only about 6 percent of rural household income derived from domestic crop production.[38]

By the mid-1930s these changes in regional political economy and the vagaries of nature had visible and dramatic effects on the Basutoland landscape. The small administrative center at Maseru had grown into a commercial center and a port of embarkation for male mine workers who were leaving their rural households for wage labor in South Africa. Mission stations, previously isolates dotted around the lowlands, had become small towns. In rural areas fields of maize stalks dominated rural textures and dongas appeared as a regular feature of disturbed soil around settlements.

In the first three years of the decade of the 1930s, drought swept southern Africa; strangely, maize crops and grain prices collapsed simultaneously, victims of drought and imported food grain, respectively. Vegetation withered, leaving topsoil unprotected. Then in 1933 torrential rains fell on the desiccated soils, and raindrop impact expanded eroded gullies around roads, paths, and cultivated fields.[39] Thus in 1935 when Alan Pim toured the countryside, he saw not only the effects of almost a century of plow agriculture but also a climatic disaster. The embattled landscape appeared to him to be the cumulative result of abuse that called for massive intervention to save it. The soil conservation program that he prescribed ensued over the next thirty years and sought to prevent further decline, but it may, in fact, have accelerated it.

Kate Showers, a soil scientist, argues that the interventions in the name of soil conservation are at the root of the modern moonscapes of dongas so visible to travelers in Lesotho in the 1990s. Actions such as meadow strips, terraces, and soil bunds that might have prevented some forms of erosion were done on the cheap and under coercion. They were often poorly designed. In many areas not yet intensively cultivated, such as her research site at ha Maghopo, the antierosion measures created dongas where none had existed before by concentrating the corrosive force of water into channels that created subsoil "piping" and then collapsed to form the new dongas that now crisscross the agricultural lowlands. Showers argues that through the 1940s and 1950s Pim's alarming words about erosion control became a self-fulfilling prophecy. His view of an impending crisis in 1935 saw an erosion crisis in the making but, like many other observers of that era, he failed to envision a local solution without formal coercion by the colonial state.

It is also true that conditions of Basutoland's rural economy and land use in place in the middle decades of this century may well have contributed to the scarred lowlands. By the mid-1930s the Basotho were only part-time farmers; most households were dependent on remittances from South African mines. Recent data confirm the decreasing role of agriculture in rural income and the overall decline of farm economies. Farms had ceased to produce for the market and relied increasingly on maize, a crop notorious for its soil depletion properties and vulnerability to drought. The plow agriculture and maize that spread out from mission station origins disturbed the fragile duplex soils far more than the Basotho's traditional hand hoes and sorghum. Under these new conditions, Basotho farms in the middle decades of this century suffered from absent male labor and a lack of traditions of soil management. When dongas appeared farms were no longer the primary source of livelihood, and declining real wages in the mines left little capital for farm investment.[40] Lesotho's massive gully erosion is thus a visible conjuncture of geology, political economy, and agricultural history.

Why is there such a striking difference between rural lands on South Africa's side of the border (the eastern Free State) and Lesotho? The answer lies as much in the impact of political economy as it does in local management and soil chemistry. In the late nineteenth century and the first third of the twentieth century there were few differences in the agricultural economies of the Free State and Lesotho. Free State farmers, whose land was more arid, developed livestock ranches, whereas the Basotho concentrated on grain production. By the 1930s and 1940s the paths of their rural economies crossed but were headed in very different trajectories. As Lesotho's rural population became more dependent on wage labor from the mines, Free State farms took advantage of a postwar boom to invest heavily in large units, cheap black labor, and the technology of dryland production of winter wheat and hybrid maize. In that period, government subsidies from South Africa's mine-based economy allowed white farms to invest in tractors, harvesters, and mechanization already tested on dryland farms in the United States and Australia.[41]

Figure 7.7 Maize cultivation and highland mesa, Lesotho, 1996. Photo by author.

Table 7.2 indicates the magnitude of wheat production for the Eastern Orange Free State in the post–World War II years.

Where dongas appeared on newly opened soils, aerial surveys spotted them and the South African government paid half the cost for farmers to build dams to convert them to livestock tanks or to construct retaining walls to hasten resiltation. The few dongas visible in the Eastern Free State today are stable or reclaimed. Lesotho, by contrast, has remained mired in the rural economy of a labor reserve: farms struggling with low labor and capital and producing primarily for subsistence. The networks of gullies that crisscross the country-

Table 7.2
**Wheat Production (in 91 kg sacks) in the Eastern Orange Free State,
1911–1969**

District	1911	1930	1949–50*	1969
Clocolan	—	—	80,946	409,828
Ficksburg	48,170	81,776	72,146	295,848
Fourisburg	—	61,683	34,657	103,575
Ladybrand	14,888	67,526	87,128	481,356
Rouxville	10,060	14,460	5,975	6,859
Smithfield	3,885	5,157	1,273	1,232
Wepener	3,459	16,322	6,916	24,039
Zastron	—	25,092	—	24,577

*Year of major drought.

Source: C. C. Eloff, *Oos-Vrystaatse Grensgordal* (Pretoria, 1980), 119.

side, collecting rusting auto carcasses and swallowing farmer's fields
are only the most visible manifestation of Lesotho's economic history.

CONCLUSION

The future of Lesotho's eroded landscape, however, may be less
bleak than many observers imagine. Alan Pim based his 1935 observations on erosion in the young, expanding stages of a growth cycle
when disturbed soils collapsed and dongas formed and grew expo-
nentially. He witnessed the outcome of several years of drought followed by major flooding without a sense that the conditions and rural
population he observed had the capacity to respond in the long term.
He and subsequent observers envisioned a trajectory spinning toward
disaster.

In fact, the life cycle of dongas viewed from the 1990s also includes
a process of stabilization in which human management can play a
significant role. Gaping dongas that appeared in the late nineteenth
century have often recovered their economic value by the late twentieth century. If dongas in their early stages expand from small cre-
vasses into yawning gashes, they change in later stages by resiltation
and collapsing walls into more stable depressions which support new
vegetation. The final stages of a donga include the growth of black

Figure 7.8 Morija donga site, c. 1915. (Courtesy of Morija Museum and Archives.)

wattle (a leguminous Australian acacia variety) that grows like a weed but stabilizes soil and provides fuel wood and forage for livestock. One major donga at Morija that was visible in a c. 1915 photograph has collapsed, resilted, and returned to viable pasture by 1996.

Farmers who can invest their labor have also discovered that a small stone wall built at the leading edge of a donga caused it to silt up rapidly, conserving moisture and building, over time, a bed of fertile soil suitable for cultivation or pasture. The Mafeteng Development Project, for example, has implemented methods of reclaiming dongas as land for gardens and domestic space.

The colonial policy that saw the appearance of urban concentrations and dongas as the harbinger of disaster may have missed a different long-term effect. A 1986 survey estimated that agriculture in Maseru and its urban periphery accounted for $2.5 million in crops and livestock. Productivity on this urban land was roughly twice that of rural land under cereal crops.[42] Thus, the urbanization that has resulted from Lesotho's inclusion in the South African industrial economy has produced a long-term side effect that may over the long term heal parts of the landscape. The growth of urban food markets has stimulated commercial agriculture and land management practices (hous-

Figure 7.9 Morija donga site eighty years later (note resiltation), 1996. Photo by author.

ing development and periurban gardens) that have halted the growth of new dongas and reclaimed others.

While many view rapid urban growth as environmentally negative, it is precisely within urban zones where land management is most developed. In Lesotho the reclamation of dongas has become a small, growing industry, particularly in areas, such as periurban zones, where land values have risen. As the urban sprawl around Lesotho's towns continues and the value of urban land increases there are and will be strong incentives for the reclamation of dongas and improved land management. How long this process may take or how far it will go into rural areas is, as yet, unclear.

NOTES

1. For their insights into Lesotho's history and documentary records, I am indebted to John and Judy Gay, Chaba Mokoko, David Ambrose, David May, Stephan Gill, Kate Showers, and Tumelo Tsikoane.

2. *Donga*, the term most commonly used in English for eroded gullies in this region, is actually a Nguni word. The Sesotho word is *lengope*.

3. A. W. Pim, *Financial and Economic Position of Basutoland: Report of the Commission Appointed by the Secretary of State for Dominion Affairs, January 1935* (London, 1935), 5, 134–39.

4. Showers, "Soil Erosion," 263–86.

5. John Gay, Debby Gill, and David Hall, eds., *Lesotho's Long Journey: Hard Choices at the Crossroads* (Maseru, 1995), 102–6.

6. Murray, *Families Divided*, 1–36. For a more biographical account of rural change, see Charles van Onselen, *The Seed Is Mine: The Life of Kas Maine, a South African Sharecropper, 1894–1985* (New York, 1996).

7. Showers, "Soil Erosion," 263–86.

8. T. M. O'C. Maggs, *Iron Age Communities of the Southern Highveld*, (Pietermaritzberg, 1976), 11–16.

9. My thanks to Kate Showers and Tumelo Tsikoane for this insight into Sesotho cultural geography.

10. For description of the hunting technique, see Thomas Arbousset, *Missionary Excursion*, ed. and trans. David Ambrose and Albert Brutsch (Morija, 1991), 72. For wheat cultivation, see E. Casalis, *Mes souvenirs* (Paris, 1930), n.p., cited in R. C. Germond, ed., *Chronicles of Basutoland: A Running Commentary on the Events of the Years 1830–1902 by the French Protestant Missionaries in Southern Africa* (Morija, 1967), 444.

11. E. Casalis, *Mes souvenirs* (Paris, 1930), n.p., cited in Germond, ed., *Chronicles of Basutoland*, 62.

12. Lesotho is the name given to their historical nation as well as the modern state by Basotho. Basutoland was the official name of the British colony separated from the Cape in 1884.

13. Elizabeth Eldredge, *A South African Kingdom: The Pursuit of Security in Nineteenth-Century Lesotho* (Cambridge, 1995), 4. See also Elizabeth Eldredge, "Drought, Famine and Disease in Nineteenth-Century Lesotho," *African Economic History* 16 (1987): 62–63. Khoisan people are often called Kora in early historical literature.

14. Arbousset, *Missionary Excursion*, 96–98. Moshoeshoe often described specific areas in terms of specific memories (see "Soil Meanings," below).

15. Casalis, *Souvenirs*, quoted in Germond, *Chronicles*, 23.

16. The Barolong, for example, arrived at Thaba Nchu under similar circumstances and later disputed Moshoeshoe's claim on them. See Colin Murray, *Black Mountain: Land, Class, and Power in the Eastern Orange Free State, 1880s to 1980s* (Washington, D.C., 1992). See also Germond, *Chronicles*, 152, for a description of an early Boer settlement at Beersheba.

17. Pim, *Financial and Economic Position*, 9.

18. Casalis, quoted in Germond, *Chronicles*, 26.

19. Setiloane cites Moffat and Willoughby's accounts of Sesotho/Setswana creation narratives. See Gabriel Setiloane, *African Theology: An Introduction* (Johannesburg, 1986), 5–6. But see Germond, *Chronicles*, 513,

for Moshoeshoe's view that the world has always existed but that animals and humans came later, animals first.

20. Germond, *Chronicles*, 455.

21. Arbousset, *Missionary Excursion*, 98. Moshoeshoe's recounting of a landscape of cultural memory is a tour de force reminiscent of Simon Schama's *Landscape and Memory* (New York, 1995).

22. Eldredge, *South African Kingdom*, 1–17.

23. Germond, *Chronicles*, 61, quoting Thomas Arbousset from June 1833.

24. Germond, *Chronicles*, 66.

25. Alice Balfour in 1895 quoted Catholic priests at Roma on Basotho aversion to planting trees around cultivated areas. See Alice Balfour, *Twelve Hundred Miles in a Waggon* (London, 1895), 44–45. Elsewhere there is evidence that farmers by 1863 had adopted planting of peach trees. See Germond, *Chronicles*, 460.

26. There are numerous observations of mid–nineteenth-century use of fire in field management. See Germond, *Chronicles*, 28, 63, 67, 454. Colonial policy seems to have restricted the use of fire in the twentieth century. John Gay has pointed out to me that fire is not used in postcolonial agriculture.

27. F. Maeder, December 1850, quoted in Germond, *Chronicles*, 55–56.

28. Leonard Thompson, *Survival in Two Worlds: Moshoeshoe of Lesotho, 1786–1870* (Oxford, 1975), 124.

29. For a fuller account of the political history of this period, see Thompson, *Survival*, 105–70; 219–77.

30. Anonymous missionary letter quoted in Germond, *Chronicles*, 319.

31. A. F. Robertson, *The Dynamics of Productive Relationships: African Share Contracts in Comparative Perspective* (Cambridge, 1987), 137; Murray, *Divided Families*, 1 ff.

32. Balfour, *Twelve Hundred Miles*, 44–45.

33. Ibid., 48.

34. Kate Showers, "Gully Erosion in Lesotho and the Development of Historical Environmental Impact Assessment," Boston University African Studies Center Working Paper No. 201, 1995, pp. 3–4; Pim, *Financial and Economic Position*, 140.

35. Robertson, *Dynamics of Productive Relationships*, 134.

36. Gay, Gill, and Hall, *Lesotho's Long Journey*, 108, describe survey data on the loss of male labor.

37. Murray, *Divided Families*, 18

38. Ibid., 19.

39. I am grateful to David Ambrose of the National University of Lesotho for climate data from the 1930s.

40. In 1931 the entry-level wage bought 18 kilograms of maize; in 1970

the entry wage rate bought only 10 kilograms. See Robertson, *Dynamics of Productive Relationships*, 134–35.

41. William Beinart and Peter Coates, *Environment and History: The Taming of Nature in the USA and South Africa* (London, 1995), 51–71.

42. Gay, Gill, and Hall, *Lesotho's Long Journey*, 51–52.

Epilogue: Africa's Environmental Future as Past

Imagine a traveler in Africa in 1800 moving across any of the landscapes described in this book. Whether in Lesotho, Ghana, Ethiopia, or the West African savanna, what would have met the viewer's gaze would differ substantially from one place to another, from Afro-Alpine highlands to forest/savanna mosaic. Yet, the agents that shaped the landscapes would have been similar: fire, rainfall, geology, dramatic seasonal change, and human tools such as plows, machetes, and steel axes. Were that traveler a pastoralist, his or her cultural imagination might focus on water points, palatable grasses, or paths that skirt tsetse-infested thickets; a farmer would notice soil density, moisture, and color; a European missionary or geographer might worry over map coordinates or imagine African versions of bucolic Flemish landscape art. In any of these cases, the landscapes would have reflected the sum of local forces, natural and human, climatological and economic.

How would the landscapes of those places present themselves to a traveler of the late twentieth century? First, a modern traveler would likely view the surroundings from a road, moving quickly in a bus, car, train, or from above in an airplane. On the landscapes themselves, geology and general patterns of climate would, then as now, act as conservative rather than dynamic forces in environmental change. Yet, it would be clear that changes in human land use and human population had wrought the most profound transformations of Africa's landscape mosaic over the two centuries of this study. Of the forces

that have shaped the interaction of humans and their physical surroundings only population has proven to be consistently cumulative
in affecting the look of the land in the late twentieth century. Yet,
even with the hindsight of history it would be difficult to predict the
condition and form of any particular place. Factors have been conjunctural or episodic as in southern Ethiopia's forest shift from open
agricultural land back to forest or the Serengeti plain conversion from
open grassland to Croton forest woodland and back. Even so, major
trends in resource use, such as deforestation or soil erosion, have depended in the historical record on migration and fertility, products of
economic and social forces that shape where people concentrated and
how they organized labor and land use.

At the same time, African landscapes have been most affected by
what an earlier generation of scholars, following Monica Wilson and
John Iliffe, called the "expansion of scale" and what is now more
generally described as political ecology and globalization, that is, the
effect of wider forces of political economy on local environmental
resources. What had been local in Africa has become, over the two-
century span of this study, increasingly regional, national, and global.
Thus, the history of Africa's environment is conjunctural, a set of
overlapping processes and events. We must therefore concentrate our
efforts to explore the evidence of specific cases and context to docu-
ment a full range of experiences and possibilities. This book has offered only a few examples of the continent's great complexity.

The effect and the sequence of this increase in scale varied between
the cases considered in this book. Lesotho's economic and environmental transformation, symbolized in the disturbance of its fragile
soils, moved early in our period and perhaps most profoundly as its
labor, agricultural produce, and economy diffused into the larger
South African economy, led by mineral wealth and international capital. The Ghanaian forest's process of change began even earlier, in the
fifteenth century, but evolved much more slowly as local invention
of forest fallow cultivation by a new and growing labor force coexisted
with its forest setting until the shifting of the Atlantic economy in the
mid–nineteenth century began to extract natural commodities (palm
oil, cocoa, and timber) rather than human souls. The subsequent landscape change, including the rise of permanent maize cultivation in the
forest biome, has only appeared in the second half of the twentieth
century. But does historical change necessarily imply degradation of
landscapes and natural resources?

Stories described here of a putative degradation in Ethiopia, southern Africa, East Africa, and the Sahel also derived ultimately from the
expansion of economic scale and the extension of international forces
onto African landscapes. In virtually all of the cases examined in this
book, alarm at the decline of Africa's natural resources was a product
of the maturation of international (i.e., colonial) interests in Africa's
economic potential between the early 1900s (in French West Africa)
and the 1930s (in Lesotho and East Africa). In a number of cases
(Ethiopia, Guinea, Machakos) assertions of degradation are misleading. In others (Lesotho, Ghana) major ecological changes have taken
place but need to be understood in a wider context.

Stories of desertification and the degradation of forests, soils, and
biodiversity had their origins in European conceptions of an African
Eden and the initial environmental effects of population expansion
and commercial extraction of agricultural and forest products. Narratives of environmental degradation are durable and persistent because they tell a plausible story about the past and the present and
offer an implied solution. Stories about forest loss or soil fertility
declining from an idealized past are not only powerful unto themselves but are embedded within broadly accepted ideas about environmental degradation as a whole and the role of local peoples,
especially rural populations, deemed responsible.

The origins of such degradation narratives have been well described
in studies by Anderson, Fairhead and Leach, Hoben, Tiffen, Mortimer, and Farah.[1] Ethiopia's degradation narratives began somewhat
later, in the post–World War II era, with the arrival of international
development efforts. In each of these cases, however, the effect has
been to set up a conflict between the global forces of conservation
and those local and international interests that are pushing for the
exploitation of natural resources as an engine for development. Both
sides of this struggle over Africa's rural resources have engaged in the
manipulation of narratives of degradation either to blame local people's mismanagement or to see rural people as passive victims.

Returning to the view of our 1800 traveler on late–twentieth-
century rural landscapes, our observer might make note that agricultural fields (whether forest fallow cultivation, swidden, or annual
cereal cropping) had greatly expanded their scale over two hundred
years. Yet, now those fields contain predominately New World crops
of maize (now Africa's dominant food crop in virtually all but the
most arid regions), groundnuts, and cassava rather than Africa's his-

torical mainstays, namely, sorghum, millet, and rice. African "forests" and woodlots as often as not display exotic non-African trees (eucalyptus, white pine, black wattle, and Asian teak) rather than local hardwoods. In open rural spaces new varieties of domestic livestock dodge cars and trucks on roads that penetrate the cities' hinterlands like the tendrils of a weed.

In the era that has followed African countries' formal independence between 1956 (Sudan), 1957 (Ghana), and 1980 (Zimbabwe), there has been another major trend affecting African environments and perceptions of them. This overwhelming process has been the centralization of economic power and the authority of the state in urban areas. In the nineteenth century Africa south of the Sahara was the world's least urbanized area: at the end of the nineteenth century only about 5 percent of Africans lived in urban areas; in 1960 only 12 percent did so. At the end of the twentieth century, Africa's level of urbanization south of the Sahara is less than half that of Latin America and the Caribbean. Yet, in the last 25 years Africa's rate of urban growth has been the fastest in the world.[2] Cities like Dar Es Salaam, Maputo, Accra, and Abidjan have begun to approach the levels of Buenos Aires, Mexico City, and São Paolo in their dominating percentage of national population. Rural land that had once produced food for urban markets is now under asphalt and periurban housing. At the same time, the fastest rates of growth and transformation in Africa have most recently been those district towns that have grown from market and administrative centers to complex urbanizing scenes, transforming their own hinterlands, economic infrastructures, food supply, and environmental settings.

The African urban process is the product of a historical process that is unique to the late twentieth century. Some of these special qualities reveal themselves idiosyncratically in still sketchy demographic data that offer many contradictions and conundrums: Africa has the world's lowest percentage of women (10 percent) in the industrial work force; Africa is among the world's least urbanized world regions, yet it ranks among the highest in percentage of each nation's urban population in its largest cities. Strangely, South Africa continues to exclude its densely settled, sprawling townships from electoral data on "urban" zones, possibly because their history as labor reserves and peculiar lack of infrastructure marks them as not "conventionally" urban.

The growth of cities and the redirected flows of natural resources, rural investment, transfers of political authority, and patterns of land use constitute a powerful force that is affecting Africa's landscapes and environmental resources. In most African cities it is far easier to phone, fax, surf the Internet, or E-mail Europe or America than to phone a relative in a rural area.

African cities are more than islands of activity and social experience; we must therefore reconsider the ways in which we understand their role in evironmental change. Demography, people's local strategies, and land practice must consider and be redefined by the complex strategies and unintended consequences of Africa's urban process. In Maseru (Lesotho) and Ethiopian towns, for example, urbanization in the long run has meant increasing tree cover and soil stabilization even if short-term effects have appeared to have had quite the opposite effect. In other cases, of course, urbanization destroys habitat and pollutes air and water.

Economic and political liberalization of the late 1980s and 1990s has further concentrated authority in urban areas by privatizing natural resources in the hands of urban-based classes. Local development efforts and attempts at participatory rural development have tended to fail in the face of these entrenched interests that have sought to claim control over all but the most marginal or remote natural resources.

What might shock our 1800 traveler most about African rural landscapes in 1999 would be not just the landscapes of high-rise glass and steel but the periurban sprawl that surrounds the city, where poverty and insecurity of tenure means that people plant fewer trees, rely on small plots of maize, and suffer poor sanitation. Such landscapes are still defining themselves both spatially and environmentally.

Change is inevitable. In the late twentieth century Africa is only catching up with the urbanization of the globe as a whole. Cities as expressions of human landscapes, as centers of political power, and points of access for international forces will as much or more than nature determine the shape of the African environmental future.

NOTES

1. Anderson, "Dust Bowl"; Fairhead and Leach, *Misreading the African Landscape*; Hoben, "Paradigms and Politics"; Mortimer, Tiffen, and Gichuki, *More People, Less Erosion*; Farah, "Natural Vegetation."

2. Josef Gugler, ed., *The Urban Transformation of the Developing World* (Oxford, 1996), and Akin Mabogunje, "Urban Planning and Post-Colonial State in Africa: A Research Overview," *African Studies Review* 33 (1990), 121–203.

Bibliography

Addis Tiruneh. "Gender Issues in Agroforestry." *Proceedings of the Second Workshop of the Land Tenure Project.* Trondheim: Centre for Environment and Development, 1994.

Almeida, Manoel de. *Some Records of Ethiopia, 1593–1646.* Translated and edited by C. W. Beckingham and G. W. B. Huntingford. London: Hakluyt Society, 1954.

Amanor, Kojo. *The New Frontier. Farmer's Response to Land Degradation: A West African Study.* London and Geneva: Zed Books, 1994.

Anderson, David. "Depression, Dust Bowl, Demography, and Drought: The Colonial State and Soil Conservation in East Africa during the 1930s." *African Affairs* 83 (1984): 321–43.

Antinori, Orazio. "Lettera del M. O. Antinori a S. E. il comm. Correnti Presidente dell Società." *Bollettino dell Società Geografica Italiana* 16 (1879): 388–403.

Arbousset, Thomas. *Missionary Excursion.* Edited and translated by David Ambrose and Albert Brutsch. Morija: Morija Archives, 1991.

Association of Ethiopian Geographers. "First Annual Conference of the Association of Ethiopian Geographers: Population, Sustainable Use of Natural Resources and Development in Ethiopia: Program and Abstracts." May 31, 1996.

Balfour, Alice. *Twelve Hundred Miles in a Waggon.* London: Edward Arnold, 1895.

Beach, D. N. *The Shona and Zimbabwe, 900–1850.* London: Heinemann, 1980.

Beinart, William. "Introduction: The Politics of Colonial Conservation." *Journal of Southern African Studies* 15, no. 2 (1989): 143–62.

Beinart, William, and Peter Coates. *Environment and History: The Taming of Nature in the USA and South Africa.* London: Routledge, 1995.

Bernatz, John Martin. *Scenes in Ethiopia.* 2 vols. Munich and London: Bradbury and Evens, 1852.

Bianchi, Gustavo. *Alla Terra dei Galla.* Milan: Società Felsinea, 1882.

Blockhus, Jill M. Mark Dillenbeck, Jeffrey A. Sayer, and Per Wegge, eds. *Conserving Biological Diversity in Managed Tropical Forests.* Gland, Switzerland, and Cambridge: International Union for Conservation of Nature and Natural Resources, 1992.

Bojo, Jan, and David Cassells. *Land Degradation and Rehabilitation in Ethiopia: A Reassessment.* AFTES Working Paper No. 17. Washington, D.C.: The World Bank, n.d.

Borelli, Jules. *Éthiopie méridionale: Journal de mon voyage aux pays Amhara, Oromo et Sidama, septembre 1885 à novembre 1888.* Paris, 1890.

Bowdich, T. E. *Mission from Cape Coast to Ashantee.* London: J. Murray, 1819.

———. *The British and French Expedition to Teembo.* Paris, 1821.

Brancaccio, L., G. Calderoni, M. Coltorti, F. Dramis, and Ogbaghebriel Berakhi. "Phases of Soil Erosion during the Holocene in the Highlands of Western Tigray: A Preliminary Report." Paper presented at the 12th International Conference on Ethiopian Studies, East Lansing, Mich., 1994.

Breitenbach, F. von. "National Forestry Development Planning: A Feasibility and Priority Study on the Example of Ethiopia." *Ethiopian Forestry Review* 3 (1962): 41–68.

Brooks, George E. "A Provisional Historical Schema for Western Africa Based on Seven Climatic Periods." *Cahiers d'Etudes Africaines* 101–102 (1986): 1–2, 43–62.

———. *Landlords and Strangers: Ecology, Society, and Trade in West Africa, 1000–1630.* Boulder, Colo.: Westview Press, 1993.

Butzer, Karl. "Rise and Fall of Axum, Ethiopia: A Geoarchaeological Interpretation." *American Antiquity* 46, no. 3 (1981): 471–95.

Buxton, Mary A. B. *Kenya Days.* London: Arnold and Company, 1927.

Cardinal, A. W. *In Ashanti and Beyond.* London and Philadelphia: n.p., 1927.

Casalis, E. *Mes souvenirs.* 6th ed. Paris: Societé des Missions Evangeliques, 1930.

Cecchi, Antonio. *Da Zeila alle Frontiere del Caffa.* 2 vols. Rome: Ermanno Loescher, 1886.

Centro Internationale por la Mejoridad de Maize y Trigo. "Ghana's Tradition Makers." *CIMMYT Today* 13 (1989): 2–10.

Cerruli, Enrico. *Etiopia Occidentale (dallo Scioa alla frontiera del Sudan):*

Note del viaggio, 1927–1928. Roma: Sindicato Italiano Arti Grafiche, 1933.

Charney, J. G. "Dynamics of Deserts and Drought in the Sahel." *Quarterly Journal of the Royal Meteorological Society* 101 (1975): 193–202.

Ciferri, Raffaele. "Primo rapporto sul caffé nell'Africa orientale italiana." *Agricolo Coloniale* 34 (1940): 135–44.

Clapham, Christopher. *Transformation and Continuity in Revolutionary Ethiopia*. Cambridge: Cambridge University Press, 1988.

Cleaver, K., and G. Schreiber. *Reversing the Spiral: The Population, Agriculture, and Environment Nexus in Sub-Saharan Africa*. Washington, D.C.: The World Bank, 1994.

Conforti, Emilio. *Impressioni agrarie su alcuni itinerari dell'altopiano etiopico*. Florence: Regio Istituto Agronomico per l'Africa Italiana, 1941.

Connah, Graham. *African Civilizations. Precolonial Cities and States in Tropical Africa: An Archaeological Perspective*. Cambridge: Cambridge University Press, 1987.

Cronon, William. "A Place for Stories: Nature, History, and Narrative," *The Journal of American History* 78 (March 1992): 1372–76.

Crosby, Alfred. *The Columbian Exchange: Biological and Cultural Consequences of 1492*. Westport, Conn.: Greenwood Press, 1972.

Crummey, Donald. "Some Precursors of Addis Ababa: Towns in Christian Ethiopia in the Eighteenth and Nineteenth Centuries." *Proceedings of the International Symposium on the Centenary of Addis Ababa*. Addis Ababa, 1987.

Daneel, M. L. "African Traditional Religion and Earthkeeping." Unpublished paper, n.d.

———. "Earthkeeping in Missiological Perspective: An African Challenge." Unpublished paper, n.d.

Daniel Gamachu. *Environment and Development in Ethiopia*. Geneva: International Institute for Relief and Development, 1988.

DeLanghe, E., R. Swennen, and D. Voylsteke. "Plantain in the Early Bantu World." *Azania* 29–30 (1994–95): 147–60.

Dodd, Jerold L. "Desertification and Degradation in Sub-Saharan Africa: The Role of Livestock," *Bioscience* 44 (1994): 28–33.

Dublin, Holly T. "Dynamics of the Serengeti-Mara Woodlands: An Historical Perspective." *Forest and Conservation History* 35, no. 4 (October 1991): 169–78.

Dupuis, J. *Journal of a Residence in Ashantee*. 2d ed. London: Frank Cass, 1924.

Eckholm, E. and L. R. Brown. *Worldwatch Paper* No. 13. Washington, D.C.: Worldwatch Institute, 1977.

Eldredge, Elizabeth. "Drought, Famine and Disease in Nineteenth-Century Lesotho." *African Economic History* 16 (1987): 62–93.

———. *A South African Kingdom: The Pursuit of Security in Nineteenth-Century Lesotho.* Cambridge: Cambridge University Press, 1995.

Elkiss, T. H. *The Quest for an African El Dorado.* Los Angeles: Crossroads Press, 1981.

Ellis, W. S. "Africa's Sahel: The Stricken Land." *National Geographic* 172 (August 1987): 141–79.

Eloff, C. C. *Oos-Vrystaatse Grensgordal.* Pretoria: Human Science Research Council, 1980.

"Ethiopia: National Report on Environment and Development." A report Prepared for the United Nations Conference on Environment and Development, Rio de Janeiro, 1992.

Fairhead, James, and Melissa Leach. *Misreading the African Landscape: Society and Ecology in a Forest-Savanna Mosaic.* Cambridge: Cambridge University Press, 1996.

———. *Reframing Deforestation: Global Analyses and Local Realities—Studies in West Africa.* London: Routledge, 1998.

Farah, Kassim O. "Natural Vegetation." In *Environmental Change and Dryland Management in Machakos District, Kenya, 1939–90: Environmental Profile,* edited by Michael Mortimer, 51–66. ODI Working Paper No. 53, December 1991.

Fattovich, Rodolfo. "Remarks on the Pre-Aksumite Period in Northern Ethiopia." *Journal of Ethiopian Studies* 23 (1990): 1–33.

———. "Archaeology and Historical Dynamics: The Case of Bieta Giyorgis (Aksum), Ethiopia." Paper presented to European Association of Archaeologists, September 1997.

Forbes, Rosita. *From Red Sea to Blue Nile: Abyssinian Adventure.* New York: Macaulay, 1925.

Ford, John. *The Role of Trypanosomiasis in African History.* Oxford: Oxford University Press, 1971.

Freeman-Grenville, G. S. P. *The East African Coast: Select Documents from the First to the Earlier Nineteenth Century.* 2nd ed. London: Collings, 1975.

Gamst, Frederick. "Peasantries and Elites without Urbanism: The Civilization of Ethiopia." *Comparative Studies in Society and History* 12 (1970): 373–92.

Gay, John, Debby Gill, and David Hall, eds. *Lesotho's Long Journey: Hard Choices at the Crossroads.* Maseru, Lesotho: Sechaba Consultants, 1995.

Germond, Robert C., ed. *Chronicles of Basutoland: A Running Commentary on the Events of the Years 1830–1902 by the French Protestant Missionaries in Southern Africa.* Morija, Lesotho: Morija Sesuto Book Depot, 1967.

Giblin, James. "Trypanosomiasis Control in African History: An Evaded Issue?" *Journal of African History* 31, no. 1 (1990): 59–80.

Gilbert, Erik. "The Zanzibar Dhow Trade: An Informal Economy on the East African Coast, 1860–1964." Ph.D. dissertation, Boston University, 1997.

Glacken, Clarence, J. *Traces on the Rhodian Shore: Nature and Culture in Western Thought from Ancient Times to the End of the Eighteenth Century.* Berkeley: University of California Press, 1967.

Glantz, Michael, ed. *Drought Follows the Plow.* Cambridge: Cambridge University Press, 1995.

Goombridge, Brian, ed. *Global Biodiversity: Status of the Earth's Living Resources.* London: Chapman and Hall, 1992.

Gore, Albert. *Earth in the Balance: Ecology and the Human Spirit.* Boston: Houghton Mifflin Company, 1992.

Graham, Douglas. "Report on the Agricultural and Land Produce of Shoa." *Journal of the Asiatic Society of Bengal* 13 (1844): 253–96.

Grove, Richard. "Scottish Missionaries, Evangelical Discourses and the Origins of Conservation Thinking in Southern Africa, 1820–1900." *Journal of Southern African Studies* 15, no. 2 (1989): 163–74.

————. *Green Imperialism: Colonial Expansion, Tropical Island Edens, and the Origins of Environmentalism, 1600–1860.* New York: Cambridge University Press, 1995.

Gryseels, Guido, and Frank Anderson. *Research on Farm and Livestock Productivity in the Central Ethiopian Highlands: Initial Results, 1977–1980.* Addis Ababa, 1983.

Gugler, Josef, ed. *The Urban Transformation of the Developing World.* Oxford: Oxford University Press, 1996.

Guyer, Jane, and Paul Richards. "The Invention of Biodiversity: Social Perspectives on the Management of Biological Variety in Africa." *Africa* 66, no. 1 (1996): 1–13.

Hall, David, and Thuso Green. *Community Forestry in Lesotho: The People's Perspective.* Morija, Lesotho: Community Forestry Programme, 1996.

Harris, W. Cornwallis. *The Highlands of Ethiopia.* 2d. ed. 3 vols. London: Longman, Brown, Green, and Longmans, 1844.

Hill, Polly. *The Migrant Cocoa Farmers of Southern Ghana: A Study in Rural Capitalism.* Cambridge: Cambridge University Press, 1963.

Hoben, Allan. "Paradigms and Politics: The Cultural Construction of Environmental Policy in Ethiopia." *World Development* 23, no. 6 (1995): 1007–21.

————. "Paradigms and Politics: The Cultural Construction of Environmental Policy in Ethiopia." African Studies Center Working Papers No. 193, African Studies Center, Boston University, 1995.

Hoppe, Kirk. "Lords of the Flies: Environmental Images and Social Engineering in British East African Sleeping Sickness Control, 1903–
1963." Ph.D. dissertation, Boston University, 1997.

Horvath, Ronald. "Addis Ababa's Eucalyptus Forest." *Journal of Ethiopian
Studies* 6 (1968): 13–19.

———. "The Wandering Capitals of Ethiopia." *Journal of African History*
10 (1969): 205–19.

Huffman, T. H. "The Rise and Fall of Great Zimbabwe." *Journal of African
History* 12 (1972): 361–62.

———. "Snakes and Birds: Expressive Space at Great Zimbabwe." Inaugural
Lecture, University of Witwatersrand, 1981.

Huffnagel, H. P. *Agriculture in Ethiopia*. Rome: Food and Agriculture Organization, 1961.

Hulme, Mike, and Mick Kelly. "Exploring the Links between Desertification
and Climate Change." *Environment* 35 (July–August 1993): 5–11, 39–
45.

Iliffe, John. "The Origins of African Population Growth." *Journal of African
History* 30 (1989): 165–69.

———. *Africa: History of a Continent*. Cambridge: Cambridge University
Press, 1995.

International Livestock Centre for Africa. *Handbook of African Livestock
Statistics*. Addis Ababa: International Livestock Centre for Africa,
1993.

International Panel of Experts Subgroup on Biodiversity. *Biological Diversity
in the Drylands of the World*. New York: United Nations, 1995.

Isenberg, C. W., and J. L. Krapf. *The Journals of Rev. Mssrs. Isenberg and
Krapf, Detailing Their Proceedings in the Kingdom of Shoa and Journeys in Other Parts of Abyssinia*. 2d. ed. London: Frank Cass, 1968.

Johnson, M. "The Populations of Asante, 1817–1921: A Reconsideration."
Asantesem: The Asante Collective Biography Project Bulletin 8 (1978):
22–28.

Johnston, Charles. *Travels in Southern Abyssinia through the Country of the
Adal to the Kingdom of Shoa during the Years 1842–43*. 2 vols. London: J. Madden, 1844.

Kahuranaga, J. "Native Grassland of the Ethiopian Highlands." *Sinet: An
Ethiopian Journal of Science* 9 (supplement) (1986): 95–104.

Kaufman, Les. "Catastrophic Change in Species-Rich Freshwater Ecosystems." *BioScience* 42, no. 11 (1992): 847.

Kingdon, Jonathan. *Island Africa: The Evolution of Africa's Rare Animals
and Plants*. Princeton: Princeton University Press, 1989.

Klein, Norman A. "Toward a New Understanding of Akan Origins." *Africa*
66, no. 2 (1996): 248–73.

Kotze, Elna, and John McAllister. "Wakkerstroom: A Working Wetland."

Paper presented to the Environmental History Workshop, University of Natal-Pietermartizberg, July 1996.

Kuzwayo, Ellen. *Call Me Woman*. San Francisco: Spinsters Ink, 1985.

Lamb, Peter J. "Large Scale Tropical Atlantic Surface Circulation Patterns Associated with Sub-Saharan Weather Anomalies." *Tellus* 39 (1978): 240–51.

Leach, Melissa, and Robin Mearns. *The Lie of the Land: Challenging Received Wisdom in African Environmental Change and Policy*. Oxford: James Currey Publishers; Portsmouth, N.H.: Heinemann, 1996.

Livingstone, F. B. "Anthropolgical Implications of the Sickle-Cell Gene Distribution in West Africa." *American Anthropologist* 60, no. 3 (1958): 533–62.

———. "Who Gave Whom Hemoglobin S: The Use of Relative Restriction Site Haploptype Variation for the Interpretation of the Evolution of the B^s-Globin Gene." *American Journal of Human Biology* 1, no. 3 (1989): 289–302.

Logan, W. E. M. *An Introduction to the Forests of Central and Southern Ethiopia*. Oxford: Oxford University Press, 1946.

Louw, W. J. "Orange Free State Rainfall. Part I: General Characteristics." Pretoria, 1979. Mimeographed.

Mabogunje, Akin. "Urban Planning and Post-Colonial State in Africa: A Research Overview."*African Studies Review* 33 (1990): 121–203.

Maggs, T. M. O'C. *Iron Age Communities of the Southern Highveld*. Pietermaritzburg, South Africa: Council of the Natal Museum, 1976.

Manning, Patrick. *Slavery and African Life: Occidental, Oriental, and African Slave Trades*. Cambridge: Cambridge University Press, 1990.

Massaja, Guglielmo. *I miei trentacinque anni di missione nell'alta Etiopia*. 12 vols. Milan: Tipografica S. Giuseppe, 1886.

May, David. *Social Forestry in Lesotho: Records of Initiatives and Achievements before the Start of the Lesotho Woodlot Project*. Maseru, Lesotho: Forestry Division, 1992.

McCann, James C. *From Poverty to Famine in Northeast Ethiopia*. Philadelphia: University of Pennsylvania Press, 1987.

———. "Children of the House." In *The End of Slavery in Africa*, edited by Suzanne Miers and Richard Roberts. Madison: University of Wisconsin Press, 1988.

———. *People of the Plow: An Agricultural History of Ethiopia, 1800–1990*. Madison: University of Wisconsin Press, 1995.

McCaskie, T. C. *State and Society in Pre-Colonial Asante*. Cambridge: Cambridge University Press, 1995.

Meggers, Betty J., Edward S. Ayensu, and W. Donald Duckworth, eds. *Tropical Forest Ecosystems in Africa and South America: A Comparative Review*. Washington, D.C.: Smithsonian Institution Press, 1973.

Meghen, C., D. E. MacHugh, and D. G. Bradley, "Genetic Characterization and West African Cattle." *World Animal Review* 78, no. 1 (1994): 59–66.

Messerli, B., and K. Aerni, eds. *Simen Mountains, Ethiopia.* Vol. 1, *Cartography and Its Application for Geographical and Ecological Problems.* Bern: Geographische Institut der Universitat Bern, 1978.

Miers, Suzanne, and Richard Roberts, eds. *The End of Slavery in Africa.* Madison: University of Wisconsin Press, 1988.

Moran, E. F. "Deforestation and Land Use in the Brazilian Amazon." *Human Ecology* 21, no. 1 (1993): 1–21.

Mordini, Antonio. "Un riparo sotto roccia con pitture rupestri nell'Ambà Focadà," *Rassegna di Studi Etiopici* 19 (1941): 54–60.

Mortimer, Michael, ed. *Environmental Change and Dryland Management in Machakos District, Kenya, 1939–90: Environmental Profile.* ODI Working Paper No. 53, December 1991.

Mortimer, Michael, Mary Tiffen, and Francis Gichuki. *More People, Less Erosion: Environmental Recovery in Kenya.* London: Wiley and Sons, 1993.

Murray, Colin. *Families Divided: The Impact of Labour in Lesotho History.* Johannesburg: Raven Press, 1981.

———. *Black Mountain: Land, Class, and Power in the Eastern Orange Free State, 1880s to 1980s.* Washington, D.C.: Smithsonian Institution Press, 1992.

Mutiso, S. K., Michael Mortimer, and Mary Tiffen. "Rainfall." In *Environmental Change and Dryland Management in Machakos District, Kenya, 1939–90: Environmental Profile*, edited by Michael Mortimer, 3–4, 13. ODI Working Paper No. 53, December 1991.

Nicholson, S. E. "Climatic Variations in the Sahel and Other Africa Regions during the Past Five Centuries." *Journal of Arid Environments* 1 (1978): 3–24.

———. "The Methodology of Historical Climate Reconstruction and Its Application to Africa." *Journal of African History* 20, no. 1 (1979): 31–49.

Nyerges, A. Endre. "Deforestation History and the Ecology of Swidden Fallows in Sierra Leone." Boston University African Studies Center Working Paper No. 185, 1994.

———. "Ethnography in the Reconstruction of African Land Use Histories: A Sierra Leone Example." *Africa* 66, no. 1 (1996): 122–44.

Otterman, J. "Baring High-Albedo Soils by Overgrazing: A Hypothesized Desertification Mechanism." *Science* 186 (1974): 531–33.

Pankhurst, Helen. "The Value of Dung." In *Ethiopia: Problems of Sustainable Development: A Conference Report.* Trondheim: Centre for Environment and Development, 1989.

———. *Gender, Development, and Identity: An Ethiopian Study*. London: Zed Press, 1992.

Pankhurst, Richard. "The Foundation and Growth of Addis Ababa to 1935." *Ethiopia Observer* 6 (1962): 33–61.

Parren, Marc P. E. "French and British Colonial Forest Policies: Past and Present Implications for Côte D'Ivoire and Ghana." Working Paper No. 188, History of Land Use Series, African Studies Center, Boston University, 1994.

Perbody, J. R. *Machakos District Gazetteer*. Machakos: Machakos District Office, Ministry of Agriculture, 1958.

Phimister, I. R. "Pre-Colonial Gold Mining in Southern Zambezia: A Reassessment." *African Social Research* 21 (1976): 1–30.

Phororo, D. R. *Livestock Farming in Lesotho and Pasture Utilization*. Maseru, Lesotho: Ministry of Agriculture, 1979.

Pim, A. W. *Financial and Economic Position of Basutoland: Report of the Commission Appointed by the Secretary of State for Dominion Affairs, January 1935*. London: His Majesty's Stationary Office, 1935.

Planning Unit, Ministry of Agriculture, Lesotho. "Brown Swiss Cattle: Draft Project Proposal." January 1979.

Posnansky, Merrick. "Prelude to Akan Civilization." In *The Golden Stool*, edited by Enid Schildkrout. New York: American Museum of Natural History, 1987.

Pyne, Stephan J. *Vestal Fire: An Environmental History, Told Through Fire, of Europe and Europe's Encounter with the World*. Seattle: University of Washington Press, 1997.

Rackham, Oliver. *Ancient Woodland: Its History, Vegetation and Uses in England*. London: Edward Arnold, 1980.

Rasmussan, Eugene. "Global Climatic Change and Variability: Effects of Drought and Desertification in Africa." In *Drought and Hunger in Africa: Denying Famine a Future*, edited by Michael Glantz, 3–22. Cambridge: Cambridge University Press, 1987.

Rege, J. E. O., G. S. Aboagye, and C. L. Tawah. "Shorthorn Cattle of West and Central Africa. I: Origin, Distribution, Classification, and Population Statistics." *World Animal Review* 78, no. 1 (1994): 2–13.

Relief and Rehabilitation Commission (Ethiopia). *Combatting the Effects of Cyclical Drought in Ethiopia*. Addis Ababa, 1985.

Robertson, A. F. *The Dynamics of Productive Relationships: African Share Contracts in Comparative Perspective*. Cambridge: Cambridge University Press, 1987.

Robinson, David. *The Holy War of Umar Tal: The Western Sudan in the Nineteenth Century*. Oxford: Clarendon Press, 1985.

Rochet d'Héricourt, Charles. *Voyage sur la côte orientale de la mer Rouge dan le pays d'Adel et le royaume de Choa.* Paris: Arthus Bertrand, 1841.

Rossel, Gerda. "Musa and Ensete in Africa: Taxonomy, Nomenclature, and Use." *Azania* 29–30 (1994–95): 130–46.

"Satellites Expose Myth of Marching Sahara." *Science News* 140 (July 1991).

Schama, Simon. *Landscape and Memory.* New York: Alfred A. Knopf, 1995.

Setiloane, Gabriel. *African Theology: An Introduction.* Johannesburg: Skotaville Publishers, 1986.

Seyfu Ketema. *Tef (Eragrostis tef): Breeding, Genetic, Resources, Utilization and Role in Ethiopian Agriculture.* Addis Ababa: Institute of Agricultural Resources, 1993.

Shaxson, T. F. "Thabana Morena Integrated Rural Development Project: Lesotho." Maseru Lesotho: 1991.

Sheddick, Vernon. "Land Tenure in Basutoland." Colonial Research Series No. 13. London, 1954.

Showers, Kate B. "Soil Erosion in the Kingdom of Lesotho: Origins and Colonial Response, 1830s–1950s." *Journal of Southern African Studies* 15, no. 2 (1989): 263–86.

———. "Early Experiences of Soil Conservation in Southern Africa: Segregated Programs and Rural Resistance." African Studies Center Working Paper No. 184, Boston University, 1994.

———. "Gully Erosion in Lesotho and the Development of Historical Environmental Impact Assessment." Boston University African Studies Center Working Paper No. 201, 1995.

———. "Soil Erosion in the Kingdom of Lesotho and Development of Historical Environmental Impact Assessment." *Ecological Applications* 6, no. 2 (1996): 653–64.

Soleillet, Paul. *Voyages en Ethiopie (January 1882–October 1884): Notes, lettres et documents divers.* Rouen: Cagniard, 1886.

Soza, R. F., B. Asafo-Adjei, S. Twumasi-Afriyie, K. O. Adu-Tutu, and B. Boa-Amponsem. "Increasing Maize Productivity in Ghana through an Integrated Research Extension Approach." Unpublished paper, Crop Research Institute, Ghana, 1996.

Stahl, Michael. "Ecology and Poverty in the Ethiopian Highlands: How Are They Connected?" Paper contributed to the Sahel Workshop, Sandbjerg Manor, Sonderberg, Denmark, January 3–5, 1996.

Summers, R. *Inyanga: Prehistoric Settlements in Southern Rhodesia.* Cambridge: Cambridge University Press, 1958.

Sumner, Claude. *Ethiopian Philosophy.* Vol 3, *The Treatise of Zar'a Ya'eqob and of Walda Heywat: An Analysis.* Addis Ababa: Addis Ababa University, 1978.

———. *Oromo Wisdom Literature.* Vol. 1, *Proverbs Collection and Analysis.* Addis Ababa: Addis Ababa University, 1995.

Sutcliffe, J. Peter. *Economic Assessment of Land Degradation in the Ethiopian Highlands: A Case Study.* Addis Ababa: Ministry of Planning and Development, 1993.

———. "Soil Conservation and Land Tenure in Highland Ethiopia." *Ethiopian Journal of Development Research* 17, no. 1 (1995): 63–87.

Swift, Jeremy. "Desertification: Narratives, Winners and Losers." In *The Lie of the Land: Challenging Received Wisdom on the African Environment,* edited by Melissa Leach and Robin Mearns, 73–90. Oxford: Oxford University Press, 1996.

Teller, Charles. "Population as a Driving Force in Agricultural and Environmental Change in Ethiopia: Implications for Agro-Ecological and Community Level Research and Education." Paper presented to the National Workshop on Environment and Population Education Network Formation in Ethiopia, 1996.

Tewolde Berhan Gebre Egziabher. "Ethiopian Vegetation: Past, Present and Future Trends." *Sinet: An Ethiopian Journal of Science* 9 (supplement) (1986): 1–14.

Thomas, D. B. "Soil Erosion." In *Environmental Change and Dryland Management in Machakos District, Kenya, 1939–90: Environmental Profile,* edited by Michael Mortimer, 27–28. ODI Working Paper No. 53, December 1991.

Thompson, Leonard. *Survival in Two Worlds: Moshoeshoe of Lesotho, 1786–1870.* Oxford: Oxford University Press, 1975.

Thornton, John. *Africa and Africans: The Making of the Atlantic World.* Cambridge: Cambridge University Press, 1996.

Thorp, Carolyn. *Kings, Commoners and Cattle at Great Zimbabwe Tradition Sites.* Harare: National Museums and Monuments of Zimbabwe, 1995.

Tiffen, Mary, Michael Mortimer, and Francis Gichuki. *More People, Less Erosion: Environmental Recovery in Kenya.* Chichester: John Wiley, 1994.

Transitional Government of Ethiopia. *National Conservation Strategy.* Vol. 1, *National Policy on the Resources Base: Its Utilization and Planning for Sustainability.* Addis Ababa: Ministry of Natural Resources Development and Environmental Protection, 1994.

Tsegaye Wodajo. *Agrometeorological Activities in Ethiopia.* Columbia, Mo: NOAA, 1984.

Tucker, Compton J., Harold E. Dregne, and Wilbur W. Newcomb. "Expansion and Contraction of the Sahara Desert from 1980 to 1990." *Science* 253 (July 1991): 299–301.

Turner, B. L., G. Hyden, and R. Kates. *Population Growth and Agricultural Change in Africa.* Gainesville: University of Florida Press, 1993.

Turner, S. D. "Soil Conservation in Lesotho." Unpublished paper, January 1975.

United Nations. *United Nations Conference on Desertification, Round-Up Plan of Action and Resolutions.* New York: United Nations, 1977.

United Nations Development Program and World Bank. *Ethiopia: Issues and Options in the Energy Sector.* Report No. 4741-ET, 1983.

United Nations Environment Program. *Status of Desertification and Implementation of the UN Plan of Action to Combat Desertification.* UNEP/GCSS.III/3. Nairobi: United Nations Environment Program, 1991.

Van Onselen, Charles. *The Seed is Mine: The Life of Kas Maine, a South African Sharecropper, 1894–1985.* New York: Hill and Wang, 1996.

Waldman, Marilyn. "The Fulani Jihad: A Reassessment." *Journal of African History* 6 (1966): 333 ff.

Waller, Richard. "Emutai: Crisis and Response in Maasailand, 1883–1902." In *The Ecology of Survival in Northeast Africa,* edited by David Anderson and Douglas Johnson. Boulder, Colo.: Westview Press, 1988.

———. "Tsetse Fly in Western Narok, Kenya." *Journal of African History* 31 (1990): 81–101.

Webb, James L. A. *Desert Frontier: Ecological and Economic Change along the Western Sahel, 1600–1850.* Madison: University of Wisconsin Press, 1995.

Wilks, Ivor. "The Northern Factor in Ashanti History: Begho and the Mande." *Journal of African History* 2, no. 1 (1961): 25ff.

———. *Asante in the Nineteenth Century: The Structure and Evolution of a Political Order.* Cambridge: Cambridge University Press, 1975.

———. "Land, Labour, Capital and the Forest Kingdom of Asante: A Model of Early Change." In *The Evolution of Social Systems,* edited by Jonathan Friedman and M. J. Rowlands. Pittsburgh: University of Pittsburgh Press, 1978.

———. "The Population of Asante, 1817–1921: A Rejoinder." *Asantesem: The Asante Collective Biography Project Bulletin* 8 (1978): 22–28 and 28–35.

———. *Forests of Gold: Essays on the Akan and the Kingdom of Asante.* Athens: Ohio University Press, 1993.

Wolde, Michael Kelecha, *A Glossary of Ethiopian Plant Names.* Addis Ababa: Addis Ababa University Press, 1987.

World Bank. "The Forest Sector: A World Bank Policy Paper." 1991.

Wrigley, C. C. "Population in African History." *Journal of African History* 20 (1979): 129–31.

Wylde, Augustus. *Modern Abyssinia.* London: Methuen, 1901.

Index

About the Author

James C. McCann is Professor of History and director of the African Studies Center at Boston University. Professor McCann has written extensively on the history of famine and the environment in Ethiopia. He has also served as consultant to Oxfam America, Oxfam (U.K.), the United Nations Environmental Programme, and the International Livestock Resources Institute.